**Recent Titles in the
Science 101 Series**

Evolution 101
Randy Moore and Janice Moore

BIOTECHNOLOGY 101

BIOTECHNOLOGY 101

BRIAN ROBERT SHMAEFSKY

SCIENCE 101

GREENWOOD PRESS
Westport, Connecticut • London

Library of Congress Cataloging-in-Publication Data

Shmaefsky, Brian.
    Biotechnology 101 / Brian Robert Shmaefsky.
       p.   cm.—(Science 101, ISSN 1931–3950)
     Includes bibliographical references (p.) and index.
     ISBN 0–313–33528–1 (alk. paper)
     1. Biotechnology.   I. Title.
  TP248.215.S56   2006
  660.6–dc22   2006024555

British Library Cataloguing in Publication Data is available.

Library of Congress Catalog Card Number: 2006024555
ISBN: 0–313–33528–1
ISSN: 1931–3950

First published in 2006

Greenwood Press, 88 Post Road West, Westport, CT 06881
An imprint of Greenwood Publishing Group, Inc.
www.greenwood.com

Printed in the United States of America

The paper used in this book complies with the Permanent Paper Standard issued by the National Information Standards Organization (Z39.48–1984).

10 9 8 7 6 5 4 3 2 1

CONTENTS

SERIES FOREWORD

What should you know about science? Because science is so central to life in the 21st century, science educators believe that it is essential that *everyone* understand the basic foundations of the most vital and far-reaching scientific disciplines. *Science 101* helps you reach that goal—this series provides readers of all abilities with an accessible summary of the ideas, people, and impacts of major fields of scientific research. The volumes in the series provide readers—whether students new to the science or just interested members of the lay public—with the essentials of a science using a minimum of jargon and mathematics. In each volume, more complicated ideas build upon simpler ones, and concepts are discussed in short, concise segments that make them more easily understood. In addition, each volume provides an easy-to-use glossary and an annotated bibliography of the most useful and accessible print and electronic resources that are currently available.

PREFACE

Biotechnology can be considered as the "automobile" of the 21st century. It is affecting almost every aspect of society in the same way as the first mass production automobile changed the world in the late 1800s. Many historians view that automobile as a phenomenal technology that brought about unparalleled global prosperity. Biotechnology is likely to bring global prosperity by providing more effective ways to grow foods, manufacture commercial products, produce energy, and treat diseases. The number of new biotechnology applications that make their way into society is increasing rapidly every year. More and more government and university laboratories are dedicating resources to biotechnology research and development. Biotechnology is becoming an increasingly popular career choice for college students enrolled in biology, chemistry, engineering, and physics programs. Many law schools offer courses and specialties in biotechnology-related areas. Allied health professionals must now receive continuing education training to understand the growing number of medical biotechnology applications they are encountering today and in the near future.

There have been considerable benefits and risks to every technology that has been introduced throughout the world in the past three centuries. For example, the automobile paved the way for rapid transportation that spurred the growth of suburbs and fast food restaurants. However, the automobile is blamed for depleted fossil fuel reserves and for considerable amounts of air pollution. The benefits of current biotechnology applications include improvements in agricultural products, safer medicines, precise treatments for genetic disorders, accurate medical diagnosis technologies, environmentally cleaner ways of producing commercial chemicals and crops, and alternatives to fossil fuels. Many of the risks include biodiversity and environmental damage caused by genetically modified organisms, unknown health risks of genetically modified foods, the potential for creating a means of inexpensive biological terrorism, and the ethic issues of cloning and gene therapy.

This book was designed to provide the reader with the basic principles of modern biotechnology. It addresses the full range of biotechnology techniques and applications used in agriculture, commercial manufacturing, consumer products, and medicine. The history of biotechnology is also covered including many of the scientists who contributed to the development of modern scientific thought and biotechnology principles. Readers are encouraged to use the unbiased information provided in this book to formulate rational opinions about the benefits and risks of biotechnology. It is also hoped that readers will appreciate the wonders of biotechnology and the creative ways in which scientists can use nature to improve human lives.

1

THE DEFINITION OF BIOTECHNOLOGY

INTRODUCTION

Biotechnology is the youngest of the sciences and is increasing in knowledge at an unprecedented rate. It is the fastest growing technical discipline and has probably gained more information per year than any other field of science. Advances in biotechnology even outpace new developments in computer science. Because of the rapid advance, biotechnology is called a revolutionary science that outpaces that ability for people to keep up with an understanding of applications in society. The term biotechnology was first used by Hungarian engineer Károly Ereky in 1919. His use of the term varies somewhat from its meaning today. Ereky used biotechnology to describe the industrial production of pigs by feeding them sugar beets as an inexpensive large-scale source of nutrients. He then generalized the term to all areas of industry in which commercial products are created from raw materials with the aid of organisms. Ereky predicted a biochemical age that rivaled the societal impacts of the Stone and Iron Ages.

The science of biotechnology is an amalgamation of biology, chemistry, computer science, physics, and mathematics. Many scientists who work in biotechnology fields have a diversity of skills that bring together two or more science disciplines. Biotechnology is also practiced as a working relationship between two or more scientists who collaborate on projects by sharing their expertise and experiences. Certain types of biotechnology involve many specialized techniques which only a few people are capable of performing. Yet, other procedures and scientific instruments used in biotechnology are fairly simple. The biotechnology concepts and techniques taught only to graduate and postdoctoral students in the 1970s are now covered in high school science classes.

Unlike earlier scientific endeavors, biotechnology relies heavily on its ability to be commercialized into a diversity of procedures and products that benefit humans. More and more scientists who enter biotechnology as a career are discovering that they need a strong business background. A great proportion of biotechnology is being practiced in industrial settings. Academic biotechnology at most universities is not carried out solely for the pursuit of information. Many of the new discoveries make their way into consumer and medical products through a process called *technology transfer*. Technology transfer is defined as the process of converting scientific findings from research laboratories into useful products by the commercial sector. The great potential for profits that biotechnological innovations can offer has changed the nature of scientific information over the past 30 years.

Scientific discoveries were once freely shared between scientists by publishing findings in professional journals. The journals were peer-reviewed meaning that other scientists familiar with the field evaluated the accuracy and validity of the information before it was published. Information in the journals was then made available through professional scientific societies and through university and industrial libraries. The advent of computer-to-computer communication systems and the Internet paved the way for inexpensive means to rapidly disseminate scientific information. Almost every new finding in biotechnology could be used to make huge profits for enterprising scientists. This started a trend in which biotechnology information is not shared freely anymore. Many scientists argue that this secrecy is stifling the progress of science and may restrict the growth of science to profit-making endeavors.

Most of the new biotechnology discoveries are patented or are protected by intellectual property rights. Patenting and intellectual property rights permit the scientists to protect their discoveries. This protection prohibits others from using the discoveries or ideas without permission or some type of payment. A patent is described as a set of exclusive rights approved by a government to a person for a fixed period of time. The patent does have a limitation in that the public has the right to know certain details of the discovery. Patents are only awarded to inventions or procedures. The person applying for a patent need not be the scientist who made the discovery. Many scientists who work for biotechnology companies are required to let the owners of the company patent the discovery.

An intellectual property right is broader in scope than a patent. It is the creation of the intellect that has commercial value. Intellectual property includes any original ideas, business methods, and industrial processes. Intellectual property rights can be granted for a lifetime. The international nature of biotechnology has led to the formation of the World Intellectual Property Organization which is located in Geneva, Switzerland. Their main goal is "to promote the protection of intellectual property throughout the world through cooperation among States and, where appropriate, in collaboration with any other international organization." A new legal term called biopiracy developed as a result of protection of biotechnology information. Biopiracy is legally interpreted as the unauthorized and uncompensated taking of biological resources.

Aside from being one of the fastest growing sciences, biotechnology is also one of the most rapidly growing industries. The U.S. Department of Labor and the President's Office of the United States have categorized biotechnology as a high-growth industry. To keep up with the rapid growth of biotechnology and its impacts on the economy, President George W. Bush in 2003 developed a set of objectives to close the workforce education gaps in the high-growth industry jobs. His goal was to have workforce training to provide people with the job skills that are needed to ensure that the changes in the global economy will not leave Americans behind. It appears that the growth of biotechnology is too fast for educators to prepare students with the current knowledge and skills needed to understand biotechnology and work in biotechnology careers.

The U.S. Department of Labor recognized the following concerns related to the growth of biotechnology careers:

- Biological technician, a key biotechnology occupation, is expected to grow by 19.4 percent between 2002 and 2012, while the occupation of biological scientists is projected to grow by 19.0 percent. (U.S. Bureau of Labor Statistics, National Employment Data)

- The biotechnology industry employed 713,000 workers in 2002 and is anticipated to employ 814,900 workers in 2007. (Economy.com, Industry Workstation, Biotech industry forecast)

- The population of companies engaged in biotechnology is dynamic and growth in the biotechnology-related workforce has been vigorous, averaging 12.3 percent annually for those companies that provided data for 2000–2002. Companies with 50–499 employees experienced the fastest growth, with an annual increase of 17.3 percent, while growth among larger firms was 6.2 percent. (U.S. Department of Commerce, A Survey of the Use of Biotechnology in U.S. Industry, Executive Summary for the Report to Congress)

Other countries are making similar assessments. Biotechnology education and training efforts are being implemented in grade schools and universities throughout Asia, Canada, Europe, and South America. Public awareness campaigns sponsored by governmental and industrial organizations are also being put in effect to keep people educated about biotechnology.

The U.S. Department of Commerce made the following observations about the global biotechnology market (U.S. Department of Commerce, Survey of the Use of Biotechnology in U.S. Industry and U.S. Bureau of Labor Statistics, 2004–05 Career Guide to Industries):

- Increasingly, companies and research organizations are seeking workers with more formalized training who have the skills of both computer and life sciences.
- For science technician jobs in the pharmaceutical and medicine manufacturing industry, most companies prefer to hire graduates from technical institutes or junior colleges or those who have completed college courses in chemistry, biology, mathematics, or engineering. Some companies, however, require science technicians to hold a bachelor's degree in a biological or chemical science.
- Because biotechnology is not one discipline but the interaction of several disciplines, the best preparation for work in biotechnology is training in a traditional biological science, such as genetics, molecular biology, biochemistry, virology, or biochemical engineering. Individuals with a scientific background and several years of industrial experience may eventually advance to managerial positions.

These conclusions are consistent with those of other nations and reflect the impacts of large technological revolutions throughout history. The invention of electrical power created a demand for new industries and updated workforce skills. Moreover, the public had to be persuaded to adopt electrical power to further fuel the growth of industries that flourished using electrical power. As recognized by the U.S. Department of Commerce, biotechnology is a broad field that requires knowledge of many sciences as well as business principles.

CONTEMPORARY DEFINITIONS OF BIOTECHNOLOGY

Most scientific terms have accurate definitions that are used strictly by the people who use science in their jobs. However, some terms such as biodiversity and biotechnology were coined by a person to mean one thing and then were interpreted to mean other things by many different people. Some of the definitions of biotechnology are narrower in scope or only address on a particular type of biotechnology. The following definitions have been used to describe biotechnology:

"The use of living things to make products." —*American Association for the Advancement of Science*

"Technologies that use living cells and/or biological molecules to solve problems and make useful products." —*Perlegen Sciences, Inc.*

"The application of the study of living things to a myriad of processes, such as agricultural production, hybrid plant development, environmental research, and much more." —*National Research Council*

"Any technological application that uses biological systems, living organisms, or derivatives thereof, to make or modify products or processes for specific use." —*World Foundation for Environment and Development*

"Biotechnology is technology based on biology, especially when used in agriculture, food science, and medicine." —*United Nations Convention on Biological Diversity*

"The application of molecular and cellular processes to solve problems, conduct research, and create goods and services." —*U.S. Commerce Department*

"The industrial application of living organisms and/or biological techniques developed through basic research. Biotechnology products include pharmaceutical compounds and research materials." —*Bio Screening Industry News*

"Applied biology directed towards problems in medicine." —*Arius Research, Inc.*

"The application of science and technology to living organisms, as well as parts, products and models thereof, to alter living or non-living materials for the production of knowledge, goods and services." —*Organisation for Economic Co-operation and Development, France*

"The ability to reliably manipulate and control living systems, from adding or subtracting a single gene to cloning an entire organism. This can be thought of as the manufacturing end of the life sciences industry." —*University of Michigan, School of Medicine*

"Body of methods and techniques that employ as tools the living cells of organisms or parts or products of those cells such as genes and enzymes." —*Lexicon Bioencyclopedia*

"Biotechnology is the integration of natural sciences and engineering sciences in order to achieve the application of organisms, cells, part thereof and molecular analogues for products and services." —*University of Hohenheim, Institute of Food Technology, Denmark*

"1. Using living organisms or their products to make or modify a substance. Techniques include recombinant DNA (see Genetic Engineering) and hybridoma technology. 2. Industrial application of biological research, particularly in fields such as recombinant DNA or gene splicing, which produces synthetic hormones or enzymes by combining genetic material from different species." —*American Foundation for AIDS Research*

"A set of biological techniques developed through basic research and now applied to research and product development. In particular, the use of recombinant DNA techniques." —*The Pew Initiative on Food and Biotechnology*

"The branch of molecular biology that studies the use of microorganisms to perform specific industrial processes." —*Princeton University WordNet*

"The use of current technologies such as DNA technologies for the modification and improvement of biological systems." —*Biotech Canada*

"Scientific process by which living things (usually plants or animals) are genetically engineered." —*EcoHealth Organization*

"A term designating the use of genetic engineering for practical purposes, notably the production of proteins in living organisms or some of their components. It is primarily associated with bacteria and mammalian cells." —*The National Centers of Competence in Research in Switzerland*

CATEGORIES OF BIOTECHNOLOGY

Biotechnology in North America is generally divided into several specialties such that each has its unique techniques and instrumentation. Agricultural biotechnology is one of the oldest areas of biotechnology and involves the production or use of domesticated animals and crops for food production. Bioenergy biotechnology is another old field of biotechnology that has been modernized into a strategy for using the metabolism of organisms to produce electricity or fuel called biofuels. Bioengineering is the use of artificially derived tissues, organs, or organ components to replace parts of the body that are damaged, lost, or malfunctioning. Bioethical biotechnology is a field of study that deals with the ethical and moral implications of biotechnology knowledge and applications. Bioinformatics is the application of artificial intelligence systems and supercomputers to handle the collection and analysis of biotechnology information.

Bionanotechnology uses biological chemicals and cell structures as the basis for microscopic computers and machines. Consumer biotechnology is involved in the use of novel biotechnology discoveries that can be used as entertainment and in household products. Diagnostic biotechnology uses biological tools to diagnose animal, human, and plant diseases. Environmental biotechnology applies the metabolism of animals, microorganisms, and plants as a means of cleaning up polluted air, soil, and water by using a strategy called bioremediation. Food biotechnology uses the metabolism of organisms to assist with the production and preservation of manufactured foods. Forensic biotechnology applies various biotechnology produces and instruments for resolving the causes and perpetrators of criminal activities.

Forest biotechnology investigates the use of microorganisms, small animals, and genetically modified plants for improving the production of commercial trees. Industrial biotechnology makes use of the metabolic reactions of organisms to produce commercially important chemicals. Marine biotechnology applies the knowledge and tools of modern biology and biotechnology to make use of, study, protect, and enhance marine and estuarine resources. Mathematical or computational biotechnology develops mathematical

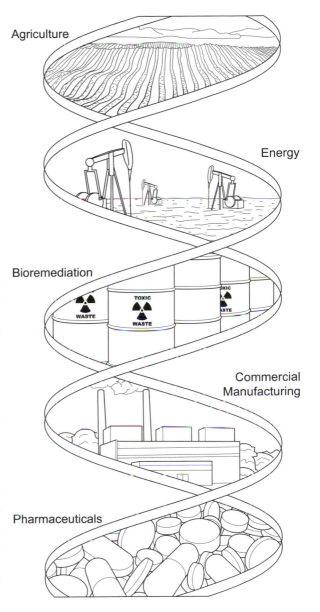

Figure 1.1 Biotechnology has many applications in agriculture, energy production, environmental sciences, manufacturing, and medicine. (*Jeff Dixon*)

and statistical formulas for interpreting biotechnology findings. Medical biotechnology looks at ways in which biotechnology produces can cure and treat human diseases. Pharmaceutical biotechnology investigates biotechnology methods for producing diagnostic materials and medications. Veterinary biotechnology deals in ways in which biotechnology produces can control and take care of animal diseases.

The European Community has developed a classification of biotechnology according to a particular industrial strategy unique to that type of biotechnology. This system of categorizing assists the various European Community nations with meeting of challenges of rapid biotechnology growth, such as job-creation and global industrial competitiveness. Each category is called a platform. Industrial platforms are a unique feature of the European Commission's biotechnology programs. Each platform is a set of technologies which are the foundation for industrial processes related to a particular type of biotechnology. All platforms have a specific mission related to the following common industrial development goals:

- Increase awareness and understanding amongst end users of the molecular techniques available and their potential applications.
- Increase awareness among technology producers of the requirements of end users.
- Provide end users with swift access to the latest technological developments and their applications.
- Develop the standard and mechanisms for training and technology transfer.
- Promote educational programs and public awareness of the role of biotechnology.

The following platforms are currently established under the guidelines of the European Community:

- ACTIP (Animal Cell Technology Industrial Platform): This platform includes animal cell technologies involved in a variety of industrial and medical applications. Some of the products of this platform include commercial proteins, hormones, medical diagnostics compounds, pharmaceutical compounds, research chemicals, and vaccines.
- LABIP (Lactic Acid Bacteria Industrial Platform): The main goal of this platform is to coordinate information and technological applications related to the genetics of the lactic acid producing bacteria. Lactic acid producing bacteria carry out many metabolic processes that have important commercial value. This platform is associated with the production of alternative fuels, dairy products, dietary supplements, industrial polymers,

and vitamins. The platform also provides a source of novel genes used in the genetic engineering of other bacteria. Another feature of this platform is bioremediation or the use of microbes to clean up contamination of air, soil, and water with pollutants.

- YIP (Yeast Industry Platform): This platform is founded on any applications of yeast-related biotechnology. A variety of yeast is used in biotechnology. However, the most commonly exploited yeast in this platform is Saccharomyces. The YIP is very important in the alcoholic beverage and food industries. Animal feeds and dietary supplements are a large part of this platform. A variety of commercial proteins, hormones, medical diagnostics compounds, pharmaceutical compounds, and research chemicals are developed in this platform.

- PIP (Plant Industry Platform): The platform is primarily involved in genetically unique plants used in agriculture, forestry, and horticulture. It also provides a source of genes used in the genetic engineering of microorganisms and plants. This platform is investigating and developing applications for the use of plants to produce commercial proteins, dietary supplements, herbal therapeutics hormones, medical diagnostics compounds, pharmaceutical compounds, research chemicals, and vaccines. Another aspect of this platform is phytoremediation or the use of plants to clean up contamination of air, soil, and water with pollutants.

- IVTIP (In Vitro Testing Industrial Platform): This platform was formed from economic, ethical, political, moral, and scientific arguments in favor of reducing or replacing the need for animal tests commonly used in medicine and research. The platform must find technologies that comply with the same governmental regulations that set the guidelines for animal testing. It involves the development of in vitro tests to reach its goal. In vitro, "in glass," refers to an artificial environment created outside a living organism which models the chemistry and functions of animals, microorganisms, and plants. The technologies used in this platform currently involve the use of animal cell cultures to replace the role of whole live animals for testing the effectiveness and safety of many consumer products. These products include chemicals such as cleaning agents, cosmetics, dietary supplements, dyes, food ingredients, fragrances, inks, preservatives, and soaps. The tests must be based on sound scientific principles and must have ample evidence to show that they provide equivalent data to animal studies.

- BACIP (Bacillus Subtilis Genome Industrial Platform): The main goal of this platform is to bring together information and technological applications related to the genetics of the Bacillus bacteria. Bacillus bacteria carry out a variety of metabolic activities that have important commercial value. This platform is associated with the production of alternative fuels, animal feeds, dietary supplements, foods, industrial polymers, and vitamins. The platform also provides a source of novel genes used in the genetic engineering of other bacteria. This platform investigates the role of Bacillus bacteria in the bioremediation of air, soil, and water.

- FAIP (Farm Animal Industrial Platform): This platform is composed of small and large agricultural operations involved in farm animal reproduction and selection. Much of the emphasis focuses on manipulating and maintaining the biodiversity of farm animals. The aim of the FAIP is to offer future lines of research on farm animal reproduction and selection to the European Community. Current applications include the genetic manipulation of domesticated animals for the production of consumer products, industrial chemicals, food, and pharmaceutical compounds. One new aspect called "pharming" uses domesticated animals that are genetically modified to produce vaccines against human infectious diseases. Other uses include the use of genetically modified animals as sources of human blood, milk, and transplant organs. The domestication of new agricultural and pet animals is also part of this platform.

- IPM (Industry Platform for Microbiology): This is a basic science platform that provides information on microbial physiology, microbial ecology, microbial taxonomy, and microbial biodiversity. It is not involved in the production of products. Rather, the IPM develops technology transfer for discoveries and research findings that have industrial applications. This platform varies greatly in the scope of microorganisms that are investigated. However, the most common microorganisms used are bacteria, fungi, and viruses. The breadth of potential produces ranges from food products to industrial chemicals.

- SBIP (Structural Biology Industrial Platform): This platform focuses more on the chemistry of organisms. It includes investigations into the structural analysis of biological molecules at every level of organization. The studies are gathered using all methods that lead to an understanding of biological function in terms of molecular and supermolecular structure. Supermolecular structure refers to the forces that cause molecules to interact with other molecules and carry out various tasks. The SBIP looks at the technology transfer potential of carbohydrates, lipid, nucleic acids, and proteins. Current products of this platform include commercial cements, industrial enzymes, medical adhesives, nanotechnology devices, preservatives, and synthetic plastics.

- BBP (Biotechnology for Biodiversity Platform): This is a basic research platform that uses information about biodiversity for technology transfer into industrial applications. Biodiversity is generally defined as the number and variety of living organisms. It takes into account the genetic diversity, species diversity, and ecological diversity of all organisms on the Earth and even on other planets. The biodiversity platform primarily focuses on the potential commercial applications of particular genes identified through biodiversity investigations. Currently, this platform identifies genes from wild plants that help crops resist diseases, drought, insects, herbicides, and poor soil quality. Cattle, poultry, and pig growers have also benefited by the discovery of genes that impart greater meat production, permit the animals to grow faster, and protect against fatal diseases. A bulk of the research conducted in this platform involves the development of genebanks. Genebanks are facilities that store the cells or DNA of all organisms on Earth. The DNA information of a genebank is also stored as a catalogue of the DNA sequence and the various traits imparted by a particular sequence of DNA.

- FIP (Fungal Industry Platform): This platform represents research and technology transfer efforts interested in biotechnology applications of filamentous fungi. Filamentous fungi or molds are microorganisms that grow as long, multicelled strands or filaments. The filaments usually come together to form larger masses such as mushrooms. This platform looks at the production of valuable molecules and materials by genetically engineered fungi. Filamentous fungi are already used in biotechnology processes used for agricultural, industrial, and medical applications. Many foods such as cheeses get their characteristic textures and flavors from filamentous fungi. Filamentous fungi also naturally produce a variety of antibiotics and pharmaceutical compounds. One group of filamentous fungi called mycorrhizal fungi is used for improving the growth of crops in poor soils. The term mychorrhae refers to the beneficial association of filamentous fungi with the small branches of roots in some plants.

- ENIP (European Neuroscience Industrial Platform): This platform focuses on medical and pharmaceutical applications related to information about the nervous system. Investigators involved in product development in this platform have produced strategies for repairing nerve damage and reversing some of the effects of stroke. This platform also deals with neural secretions that can serve as new pharmaceutical treatments for psychological disorders. Stem cell research is commonly done in the ENIP.

- EBIP (Environmental Biotechnology Industrial Platform): This is one of the newer platforms and is engaged in the field of environmental biotechnology. Environmental biotechnology is a broad field that includes a wide variety of agricultural and industrial applications. The EBIP includes the deliberate use of biological means to conserve or change the chemistry of the atmosphere, land, and water. Some current applications include soil and sediment remediation, water purification, the removal of organic and inorganic pollutants, the breakdown or biodegradation of organic pollutants, introduction of natural or genetically modified organisms to treat solid wastes, water treatment, marine cleanup, and the conversion of wastes into other materials and energy sources.

- TSE IP (TSE Industrial Platform): This platform deals with research related to transmissible spongiform encephalopathies. Transmissible spongiform encephalopathies or TSEs are fatal, incurable degenerative diseases of the brain transmitted by living agents called prions. Prions are infectious agents that are composed only of protein. TSEs are transmitted from one animal to another and produce changes in the brain which give the brain the appearance of a sponge. Mental and physical abilities deteriorate the brain and cause the formation of many tiny holes that can be seen under the microscope. The most well known TSE is called mad cow disease. However, horses, pigs, and sheep develop a similar condition. Humans also have TSEs and can get them from eating infected foods. Thus, the TSE IP uses scientific results and their applications within industry to provide the best and safest meat products possible.
- HAE 2000 (Healthy Ageing Europe Industrial Platform): This platform combines research on human aging with biotechnology innovations that may reduce ailments and diseases attributed to age. It was formed out of the need to address aging as a factor of social and economic challenges that develop in a society as people age. Research derived from this platform focuses on the preventive methods and therapies using biotechnology applications that reduce the damaging effects of aging. It involves the development of diets containing functional foods, nutritional supplements, and vaccines. Functional foods are beverages and foods claimed to have specific health benefits based on scientific evidence. These health benefits are derived from one or more nutrients or nonnutrient substances that might impart health benefits. It is hoped many of these compounds can be introduced into the foods using genetic technology and other biotechnology applications.

Another method of compartmentalizing biotechnology is on the basis of the biological principles applied in the research or processes. The accepted major kind of biotechnology categories are genomics, proteomics, metabolomics, cellomics, physiomics, and environomics. Each of these investigations as listed in their order of appearance in the previous sentence represents an increase in biological complexity. Genomics looks at the DNA level whereas environomics looks at all the environmental factors that affect an organism. There is debate about the origins of these terms. As with the term biotechnology, these terms were coined by individuals and then took on specific meanings that were accepted by the scientific community. However, they became commonly accepted by the scientific community in the late 1980s and early 1990s. Each of these categories has a particular type of knowledge, skills, and outcomes that make them career specialties and the basis of biotechnology industries.

The study of genomics is commonly categorized in chromatinomics, chromonomics, epigenomics, and ethnogenomics. Chromatinomics studies the chemistry controlling the genetic regulation of the functional DNA within a cell. Chromatin, or the functional DNA, is the substance that makes up a chromosome. It consists of pure DNA in bacteria and is an arrangement of DNA and proteins in the complex cells of higher organisms such as animals and plants. Chromatinomics is an important aspect of stem cell research. It provides the information needed to understand how the activities of a cell can be controlled by artificially manipulating the DNA. Stem cell researchers are interested in chromatinomics because it provides the ability to use stem cells as a method for healing or replacing damaged tissues. The term is used according to the definition coined by Jan Cerny and Peter J. Quesenberry in 2004 in a study titled "Chromatin remodeling and stem cell theory of relativity" published in the *Journal of Cell Physiology*.

Chromonomics is similar to chromatinomics in that it investigates DNA function. However, chromonomics differs in that it deals with the significance of gene location and arrangement on the chromosomes. Scientists use the term three-dimensional position when referring to the location and position of genes. Chromonomics research studies the influence a gene has on the function of nearby genes. In addition, it helps scientists better understand the diseases and life spans of cells, tissues, organs, and individuals. This information is also very useful for understanding the full effects of genetic manipulation on individual cells and whole organisms. The accepted use of chromonomics is found in the research of Uwe Claussen published in 2005 in the journal *Cytogenetic and Genome Research.*

Epigenomics is the science of epigenetics. Epigenetics is the study of the changes in gene regulation and traits that occur without changes in the genes themselves. It investigates any factor that affects the usage of DNA from one generation to the next. Research on epigenomics primarily focuses on the chain of command of genes in embryonic development, the development of stem cells in adult and fetal tissues, and the mechanisms of gene activation in cancer. Biotechnology makes use of epigenomics for developing therapies that aim at switching genes on and off as an approach to the treatment of aging, inherited diseases, and cancer. The accepted definition of the term first appeared in the publication "From genomics to epigenomics" in *Nature Biotechnology* written by Stephan Beck, Alexander Olek, and Jörn Walter in 1999. Mitogenomics is a type of epigenomics because it investigates the application of the complete mitochondrial genomic sequence. Other organelles such as the chloroplasts of plants also have DNA that is important to epigenetics.

Ethnogenomics, as implied in the name, evaluates the influence of ethnicity of the genomics of organisms. Ethnicity refers to organisms with origins from different parts of the world. Most scientists focus on the ethnogenomics of humans. This means that they study the characteristics of the genomic diversity found amongst various groups of populations identified as races or ethnic groups. Ethnogenomics helps medical researchers understand the racial factors that influence the distribution of genetic disorders. For example, sickle cell anemia is most prevalent in people of African and Mediterranean origin while cystic fibrosis is more common in people of northern and eastern European ancestry. Ethnogenomics has given birth to a new area of pharmaceutical biotechnology called pharmacogenomics. Pharmacogenomics is an understanding of the relationship between a person's genetic makeup and its response to drug treatment. Some drugs work well in one ethnic group and not as well in others. Biotechnology uses pharmacogenomics as the basis of designing therapeutic treatments that work more effectively without causing severe side effects. The common usage of ethnogenomics appeared in "The ethnogenomics and genetic history of eastern European peoples" published in 2003 by Elza K. Khusnutdinova in the *Herald of The Russian Academy of Sciences.*

Proteomics, or proteogenomics, goes beyond the study of the genetic material and investigates proteins programmed by the DNA. It is defined as the study of the structure and function of proteins, including the way they function and interact with each other inside cells. Stephen M. Beverley and his colleagues first used the term proteomics in their publication "Putting the Leishmania genome to work: Functional genomics by transposon trapping and expression profiling" in the Mitsubishi Kagaku Institute of Life Sciences (MITILS) of Japan 2001 Annual Report. Many researchers in biotechnology prefer to work with proteomics because it represents how the cells carry out their jobs after being genetically modified. Proteomics is a branch of transcriptomics that investigates only the proteins that is made by the DNA at a particular time or under specific conditions. The term transcriptome was used first by Victor E. Velculescu and his team in his research titled "Characterization of the yeast transcriptome" in the journal *Cell* in 1997.

Proteomics can be subcategorized into specialties such as allergenomics and enzymomics. Allergenomics focuses on the proteins involved in the immune response of animals and humans. It is derived from the term allergen. An allergen is any substance capable of inducing an allergic reaction in an animal or a person. Medical doctors describe an allergic reaction as an overreaction of the body's immune system when a person is exposed to allergens to which it is sensitive. Extreme responses to allergen are called allergies or hypersensitivities. Allergenomics is very important in the biotechnology development of diagnostic procedures, pharmaceutical compounds, and vaccines for medical and veterinary use. The word allergenomics was proposed as a standard biotechnology term in 2005 by the Division of Medical Devices, National Institute of Health Sciences in Japan.

Enzymomics is a branch of proteomics that investigates the functions of enzymes. Enzymes are complex proteins that help make a specific chemical reaction occur. Many enzymes carry out their functions inside of the cell. Other enzymes are secreted and perform a variety of jobs in body fluids or outside of the body. The categorization of an organism's enzymes is called the enzymome. This concept was first proposed in 1999 by Mark R. Martzen at the University of Rochester School of Medicine in Rochester, NY. The term enzymomics was used by Marc Vidal in an article titled "A biological atlas of functional maps" in the journal *Cell* published in 2001. Enzymomics is probably one of the fastest growing areas of industrial biotechnology. Enzymes have many applications in the production of foods, medicines, and commercial chemicals. Even enzyomomics has subcategories such as kinomics which investigates enzymes called kinases that control cell function.

Metabolomics investigates the genetics involved in the production and regulation of enzymes making up an organism's metabolism. Metabolism is best defined as the sum of the physical and chemical changes that take place in the cells of living organisms. Biotechnology applications of metabolomics primarily involve the metabolic control and regulation of the intact cells grown in cultures. Metabolomic research is important for understanding the functions of genetically modified organisms and the effects or therapeutic treatments on animals and humans. Medical researchers need metabolomic information to better understand the basics of genetic and infectious diseases. Some researchers are developing tools called microarrays that could rapidly measure the metabolomics of an organism under a variety of environmental conditions. Metabolomics was first used by Jeremy K. Nicholson and his colleagues in "'Metabonomics': Understanding the metabolic responses of living systems to pathophysiological stimuli via multivariate statistical analysis of biological NMR spectroscopic data" published in 1999 in the journal *Xenobiotica.*

Two subcategories of metabolomics are CHOmics and lipidomics. CHOmics was a term coined to describe the role of carbohydrates in metabolomics. The CHO of CHOmics is a scientific shortcut for the major carbohydrates commonly involved in animal and plant metabolism. The letter C stands for carbon, H for hydrogen, and O for the oxygen that makes up the chemistry of most carbohydrates. Scientists are learning more and more that carbohydrates play very important roles in the regulation of cells. It has recently been shown that simple biotechnology modifications of carbohydrates can be done to prevent the rejection of organs during a transplant. The term was first used by Manel Esteller in 2000 in the *New England Journal of Medicine*.

As evident in its name, lipidomics is a rapidly growing area of biotechnology in which a variety of techniques are used to understand the hundreds of distinct lipids in cells. Scientists who study lipidomics are interested in determining the molecular mechanisms through which lipids assist metabolism. Lipidomic research is currently focused on the metabolic basis of diseases in a variety of organisms. It will eventually yield new types of biotechnology products for commercial and therapeutic use. The term was first used by Xianlin Han and Richard W. Gross in "Global analyses of cellular lipidomes directly from crude extracts of biological samples by ESI mass spectrometry: A bridge to Lipidomics" in the *Journal of Lipid Research* published in 2003.

Cellomics investigates the cellome which is the entire accompaniment of molecules and their interactions within a cell. It involves studying all of the information within the cell that defines the sequence and arrangement of molecular interactions that carry out normal and abnormal functions. It represents one level of complexity above metabolomics because it factors in how the cell modifies metabolism in response to the environment and to interactions with other cells. Much of cellomics focuses on cell function during disease and impacts of drugs at the level of the cell. The term was first used in 2000 by Eugene Russo in the publication "Merging IT and biology" in the journal *The Scientist*.

Physiomics and the related science physiogenomics use the knowledge of the complete physiology of an organism, including all interacting metabolic pathway. It is a biotechnology application of physiology which is defined as the study of the overall functions of living organisms. Physiomics takes into account how the cellomics of particular body cells interact with the whole body. Currently, this area of biotechnology has focused on an understanding of the genetic basis of fundamental chemical pathways that operate the heart, lung, kidney, and blood vessels. The information is used to better diagnose and understand diseases as well as the development of biotechnology therapies. Physiome, which is the basis of physiomics and physiogenomics, was coined by James B. Bassingthwaighte at the University of Washington in 2000. Environomics investigates the effects of environmental factors on the physiome. It was developed by James C. Anthony at Michigan State University School of Medicine to describe his investigations in the genetics of environmental adaptations.

There are also overarching areas of genomic studies that use physiomic and environomic information. Behaviouromics, or the Mental Map Project, was developed by Darryl R. J. Macer of the Eubios Ethics Institute in Thailand. Research on the behaviourome currently focuses on mapping the genetics behind the sum of ideas human beings can have relating to moral decision making. Behaviouromics may ultimately branch out into research studies using biotechnology to correct behavioral disorders. Embryogenomics investigates the genes involved in the development of organisms from the point of fertilization until birth. It is a category of developmental genomics that is associated with the genetics of maturation and aging. Embryogenomics was coined in 2001 by Minoru S. Ko in "Embryogenomics: Developmental biology meets genomics" in the journal *Trends in Biotechnology*.

Biomics was established in 2002 at the Erasmus Center for Biomics in the Netherlands. It coordinates the knowledge of genomics, proteomics, and bioinformatics to develop a rational model for understanding the full functions of an organism's genetic material. Bioinformatics is the collection, organization, and analysis of large amounts of biological data, using networks of computers and databases. Bibliomics comes from the term "biblio" or book. It is a specialized aspect of biomics that investigates and applies high-quality and rare information, retrieved and organized by a systematic gathering of the scientific literature. Bibliomics uses sophisticated computer searching tools from existing databases and links all of the other biotechnology areas. It is the research focus of Bertrand Rihn's research team at the Institut National de Recherche et de S curit in France since 2003. The group is currently focusing on identifying all the research linking gene regulation to animal and human tumors.

2

BASIC SCIENCE OF BIOTECHNOLOGY

CHEMISTRY AND PHYSICS OF BIOTECHNOLOGY

Much of biotechnology takes advantage of the agricultural, commercial, and medical applications of biological molecules. Biological molecules are also called biochemicals or macromolecules. The term macromolecules stands for "macro" or large molecules because they are usually composed of many elements. Biologically, macromolecules belong to a category of molecules that chemists call organic molecules. An organic molecule is any of a large group of chemical compounds that contain carbon and are derived from organisms. Organic molecules are composed of a carbon skeleton and arrangements of elements called functional groups. Functional groups provide the molecules with their chemical and physical properties. Scientists rely on their knowledge to control the cellular processes that build biological molecules. They can modify cells' functions that build the molecules or they can carry out chemical reactions that synthesize molecules similar to those found in nature.

Many biological molecules have an important physical property called chirality. Chirality is defined as the ability of a molecule to exist in two mirror-image forms. These forms are called the left and right orientations because one type rotates polarized light in a direction opposite to the other. Chirality is determined by shining a beam of polarized light through a solution of the molecules. Polarized light is a beam of light in which the waves are all vibrating in one plane. Most organisms can only produce the same chiral form of a particular molecule. Similarly, the metabolic reactions of almost all organisms can only make use of one chiral form. For example, the glucose molecule used as a source of energy for almost all organisms is synthesized in organisms as the

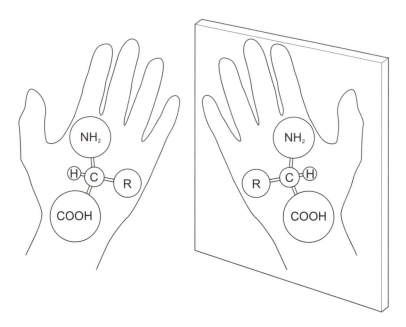

Figure 2.1 Many molecules have a property called chirality or mirror image structures. Organisms use one form or another in metabolism. One form is useful while the other form can be toxic. Certain biotechnology applications use toxic chiral forms as medicines. (*Jeff Dixon*)

"right-handed" form. The right-handed form is the only form that can be used to produce cell energy.

Chirality is important to biotechnology researchers because the correct chiral forms of a molecule are essential to growing and maintaining organisms used in biotechnology applications. Certain biotechnology procedures rely on the fact that the incorrect chiral forms can be used as therapeutic agents or as chemicals that modify the metabolism of an organism. Chirality belongs to a broader category of organic molecule properties called isomerism. Isomers are defined as molecules having the same chemical formula and often with the same kinds of bonds between atoms but in which the atoms are arranged differently. Many isomers share similar if not identical properties in most chemical contexts. Biotechnology researchers have learned to create novel biological molecules by directing an organism's metabolism to produce isomers not normally synthesized by a cell. These novel molecules can be used for a variety of purposes including glues, inks, and therapeutic compounds.

All biological molecules obey the natural laws of biophysics. Biophysics is the application and understanding of physical principles to the study of the functions and structures of living organisms and the mechanics of life processes. Scientists who study biophysics investigate the principles underlying the ways organisms use molecules to carry out living processes. The specific molecules involved in a biological process are identified using a variety of instruments and techniques used for chemical and biochemical analysis. These instruments and techniques are capable of monitoring the properties or the movement of specific groups of molecules involved in cell activities. Moreover, researchers can view and manipulate single molecules. Biotechnology applications are dependent on the relationship between biological function and molecular structure. Biophysicists can use this relationship to create precision molecules that produce predictable changes in an organism or have accurate commercial properties.

Biological thermodynamics is also an important principle for understanding the function of biological molecules in an organism. Thermodynamics is described as the relationships between heat and other physical properties such as atmospheric pressure and temperature. It comes from the Greek terms *thermos* meaning heat and *dynam* meaning power. Biological thermodynamics may be defined as the quantitative study of the energy transformations that occur in and between living organisms, body components, and cells. Quantitative study refers to observations that involve measurements that have numeric values. The measurement of thermodynamics permits biologists to explain the energy transformations that organisms carry out to maintain their living properties. Two important principles of thermodynamics that control living processes are (1) the total energy of the universe is constant and energy can neither be made nor destroyed and (2) the distribution of energy in the universe over time proceeds from a state of order to a state of disorder or entropy.

Biotechnology researchers recognize that organisms require strict chemical and physical factors in the environment for performing the work—to stay alive, grow, and reproduce. This is particularly important when they have to control the growing conditions of cells or organisms raised in laboratory conditions. An organism's ability to exploit energy from a diversity of metabolic pathways in a manner that produces biological work is a fundamental property of all living things. In biotechnology research the amount of energy capable of doing work during a chemical reaction is measured quantitatively by the change in a measurement called Gibbs free energy. Gibbs free energy, which is measured as the unit of heat called the calorie, can be viewed as the tendency of a chemical change to occur on its own accord. Organisms take advantage of nutrients which fuel the chemical reactions that give off free energy as a means of obtaining energy from the environment. This energy is then used to maintain the organism's functions and structure. Biotechnology researchers must provide organisms with molecules that maximize the energy needs. Biological thermodynamics helps biotechnology researchers predict the cell functions such as DNA binding, enzyme activity, membrane diffusion, and molecular decay. Biological thermodynamics is often called bioenergetics when used to explain energy-producing metabolic pathways.

Scientists who work in biotechnology categorize biological molecules into four fundamental groups. Each group is defined by a basic unit of structure called a monomer. A monomer is defined as a single molecular entity that may combine with other molecules to form more complex structures. One type of complex structure is the polymer. Monomers are the starting material or single unit from which a polymer is built. Polymers are defined as natural or synthetic material formed by combining monomer units into straight or branched chains. The monomers are held together by strong chemical bonds called covalent bonds. A covalent bond is formed by the combination of two or more atoms by sharing electrons. This type of bond provides the chemical stability that organisms need to survive under a variety of environmental conditions. Another type of complex structure is called the conjugated molecule. Conjugated molecules are a mixture of two or more categories of monomers or polymers bonded together to form a simple functional unit. The components of a conjugated molecule can be held together with various types of chemical bonds.

The four categories of biological molecules are carbohydrates, lipids, peptides, and nucleic acids. Carbohydrates are compounds of carbon, hydrogen, and oxygen with a ratio of two hydrogen atoms for every oxygen atom. The name carbohydrate means "watered carbon" or carbon atoms bonded to water molecules. Carbohydrates, used by all organisms as a source of nutrients for energy and body components, are synthesized by the photosynthesis carried out in plants. Monomers of carbohydrates, which are called monosaccharides, generally provide energy to living cells. Glucose and fructose are the two most common carbohydrates used for cell energy. A precise amount of these molecules in a balanced diet is necessary for maintaining the health of cells and whole organisms grown for research and biotechnology applications.

Carbohydrates also take the form of disaccharides, two different or similar monosaccharides bonded together, and polymers called polysaccharides. Disaccharides are important in biotechnology because they are commonly used for a variety of purposes including animal feeds, cosmetics, glues, and pharmaceutical compounds. Certain natural and artificial disaccharides produced by biotechnology processes are used as low-calorie sweeteners. Disaccharides are a common source of energy for the biotechnology production of biofuels. Some biotechnology companies specialize in producing natural and artificial polysaccharides for commercial purposes. Polysaccharides are integral components of thickening agents used in many absorbent materials, building materials, cosmetics, desserts, glues, paints, and pills. Several kinds of biodegradable plastics are made from polymers that decay when eaten by microbes in the environment.

Lipids, like carbohydrates, are composed primarily of carbon, hydrogen, and oxygen. Their structure is very rich in carbon and hydrogen and are often referred as hydrocarbons. Lipids, which are sometimes called fats, are categorized according to their degree of chemical complexity. Three major groups of lipids are the glycerides, sterols, and terpenes. Glyercides are composed of a fatty acid attached to a glycerol molecule. Certain glycerides called phospholipids contain the element phosphorus and are important in adapting cell structure to environmental conditions. A fatty acid is a molecule consisting of carbon and hydrogen atoms bonded in a chainlike structure. The chains of most organisms have fatty acids that range from 6 to 28 carbons in length. A glycerol molecule can bind to one, two, or three fatty acids. Monoglycerides are composed of one fatty acid chain attached to the glycerol. These lipids are very important nutrients for cells and organisms.

Diglycerides are common fats that make up cell structure. As their name implies they consist of fatty acids bonded to the glycerol. Natural and artificial diglycerides have many purposes in commercial chemical production. Triglycerides are usually composed of a glycerol molecule with three fatty acid molecules attached to it. They are usually referred to as storage fats because animals and many plants store excess calories in triglycerides. Triglycerides are used to thicken and stabilize many biotechnology products. The chemical stability of glycerides is determined by the nature of the fatty acid. Saturated fatty acids have carbons that are attached to each other by single bonds and have the maximum amount of hydrogen atoms bonded to the molecule. These fats are stable and do not readily decay. However, too many of these lipids in the diet may cause health problems in humans. Unsaturated fats are unstable and decay over time because they have fragile double bonds between some carbon atoms that are deficient in hydrogen atoms. These fats are commonly used as preservatives in biotechnology operations because they absorb any damage from environmental factors that break chemical bonds. Damage to the lipid slows down the damage to other molecules.

Sterols are a group of lipids that are similar to cholesterol in composition. They consist of a chain of carbons twisted into a pattern of rings. The hormones cortisone, estrogen, and testosterone are a type of sterol called steroids. Sterols can be synthesized in the cell from any other biological molecule. Many biotechnology researchers exploit a cell's ability to make a variety of sterols through metabolic engineering. These synthetic sterols are used in many therapeutic applications. Terpenes are a diverse group of complex fats that include hormones, immune system chemicals, and vitamins. They are also commonly synthesized in toxins and thick sticky fluids in many plants. Terpenes have many commercial applications and are a focus for many biotechnology applications. Terpene derivatives can be found in dyes, paints, pesticides, plastics, and medicines.

Peptides are often referred to as the building materials of living cells. Their elemental chemistry consists of carbon, hydrogen, and oxygen like the carbohydrates and lipids. However, they also contain nitrogen and sulfur. Proteins are the most common type of peptides found in living organisms. These molecules are often very large and are made up of hundreds to thousands of monomers called amino acids. Amino acids are a large class of nitrogen-containing organic molecules that readily form polymers using a special covalent bond called the peptide bond. Most organisms on Earth make use of approximately twenty types of amino acids that are combined in different ways to make up the one million or so different proteins. Many of these proteins contribute to cell and body structure. Others carry out chemical reactions for the organism. These proteins are called enzymes.

All of an organism's proteins are programmed for in the genetic material. The genetic material stores the information a cell needs to put together the sequence of amino acids of its various proteins. Proteins are probably the most common biological molecules for biotechnology applications. An organism's characteristics can be altered to produce desirable traits by modifying the genetic material that programs for proteins. Enzymes in particular have much commercial value because

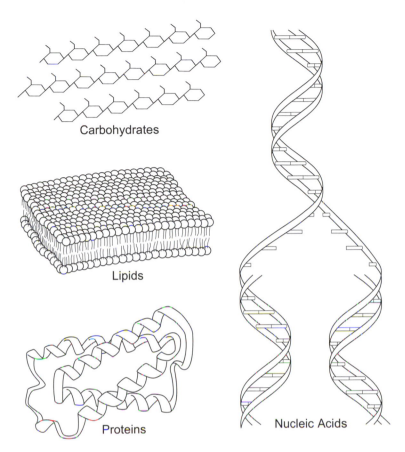

Carbohydrates

Lipids

Proteins

Nucleic Acids

Figure 2.2 Biologists categorize the molecules of living organisms into carbohydrates, lipids, proteins, and nucleic acids. (*Jeff Dixon*)

they can be used to carry out many chemical reactions used in food production, industry, and medicine. An almost unlimited variation of proteins can be synthesized using simple biotechnology procedures. In addition, it is possible to make novel proteins by adding amino acids not normally used by a living organism.

Nucleic acids are chemicals composed of a basic unit called the nucleotide. Each different type of nucleotide has a group of phosphate molecules, a monosaccharide, and a unique chemical called the nitrogen base. Nucleic acids control the processes of heredity by which cells and organisms reproduce proteins. Deoxyribonucleic acid, or DNA, is a polymer of nucleotides that contain a deoxyribose monosaccharide. Ribonucleic acid, or RNA, is another of the polymer nucleic acids. It consists of a ribose monosaccharide. There are five common types of nucleotide bases used by living organisms: adenine, cytosine, guanine, thymine, and uracil. Adenine, cytosine, and guanine are found in DNA thymine. RNA is made up of adenine, cytosine, guanine, and uracil. Uracil in RNA replaces the role of thymine which is found only in DNA. The type, location, and sequencing of the nucleotides govern the biological role of the nucleic acid. Simple nucleic acids, such as adenosine triphosphate (ATP), are involved in energy usage by cells. The role of nucleic acids in carrying out an organism's genetic characteristics is of primary importance to all biotechnology investigations and applications.

BASIC BIOLOGY OF BIOTECHNOLOGY

The basic principles of the biological sciences form the foundation for all biotechnology research and applications. Biology is coined from the Greek words *bios*, which means life, and *logos*, which means the reasoning behind or philosophy of a subject. Many people interpret biology as the study of life. Biology is concerned with the characteristics and behaviors of organisms. It deals with the mechanisms of existence of individual organisms and populations of organisms and their interaction with each other and with their environment. Biology consists of an expansive range of research fields that are often viewed as independent investigations but work with each other to build a better understanding of organisms. Many biologists incorporate science disciplines into their work as well as other fields of study such as anthropology, philosophy, psychology, and sociology.

The "life" part of biology's definition is not as simple a concept as one would imagine. Biologists generally define life with a common usage or working definition. A working definition is best described as a simple explanation encompassing most aspects or examples of the concept. A majority of biology books would provide a general working definition description such as, "life is the ongoing process of organic chemical occurrences by which living things are distinguished from nonliving ones." This definition takes into account simple organisms as well as complex ones such as humans or trees. Other books describe life as a list of characteristics that distinguish living organisms from inanimate objects. These properties comprise the following features:

- Living things obey the laws of physics and chemistry
- Living things are highly organized structures composed of organic molecules

- Living things metabolize or possess metabolic pathways that process nutrients and produce wastes
- Living things have homeostasis or the ability to self-adjust using metabolic regulation
- Living things respond and adapt to environmental changes
- Living things grow and develop
- Living things self-replicate or reproduce
- Living things have heritable material such as DNA
- Living things communicate with the environment or other living things
- Living things have some type of movement or animation
- Living things have an evolutionary origin from a single primordial life form

All of these properties describe the "typical" living organism and are somewhat biased to the characteristics exhibited by humans and related organisms.

Unfortunately, most definitions and descriptions of living things lack the sufficient conditions that enable scientists to specify whether something is living or not. For example, while metabolism is a necessary condition for living, it is by itself not a sufficient condition. This means that the presence of metabolism alone is not fully sufficient to describe living things. A living thing that shows metabolism could not survive without some of the other conditions such as the ability to adapt to the environment or the need to grow and develop. For example, certain microorganisms such as bacteria called rickettsia lack the ability to self-adjust using metabolic regulation. They have to obtain this property by living as parasites within the cells of other living things.

Some organisms lack almost all the characteristics of life and do not even fit within most definitions of life. Viruses, for example, barely meet the criteria of living things. They have a very simple structure and do not carry out any metabolic processes. In addition, they cannot even replicate without the assistance of other living things. As a result, biologists have to categorize viruses based on the characteristics they possess while infecting another living thing. It is then that viruses are able to pass along heritable material, replicate, and adapt to environmental change. Viruses were once thought to be complex life forms that forfeited many of their characteristics over time as they lived off the resources of organisms. They remain very successful organisms as long as other living things are around to provide viruses with these resources. Influenza and smallpox are examples of viruses.

Some disease-causing "organisms" completely defy the contemporary definitions of life. These purported life forms are given the designation "particles" because they do not fit even the minimum definition of life. A particle is a chemical that takes on reproductive capabilities when given the resources of a living organism. Viroids are infectious particles composed completely of a single piece of circular RNA. Ribonucleic acid is one type of heritable material that is used to pass along the characteristics of a living thing. Viroids will only replicate when an organism that they infect creates copies of the viroid's RNA. The only evidence that they are somewhat of a living thing is the presence of heritable material. Otherwise, they would not be identified as living if their chemistry was studied without knowing the consequences of placing them in another living thing. Hepatitis D, which causes liver damage and cancer, is the only human disease known to be caused by a viroid. Viroids mostly cause plant diseases.

One type of particle lacks what almost all biologists would debate is heritable material. Prions are a group of infectious particles composed exclusively of a single small protein called a sialoglycoprotein. Sialoglycoprotein resembles the proteins that help the body's immune system to identify disease-causing organisms. Prions contain no nucleic acid. This means that they have nothing traditionally recognized as heritable material. Their replication challenges the standard meaning of reproduction. Prions replicate by modifying the proteins of another organism. The organism's proteins are converted into new prions that then accumulate in the cells as a clump of prion proteins called an amyloid. The amyloid eventually kills the cell and releases the prion proteins for another round of infection and killing. Prions are associated with a variety of human diseases such as Alzheimer's disease, Creutzfeldt-Jakob disease, Down's syndrome, fatal familial insomnia, and kuru leprosy. Mad cow disease, or bovine transmissible spongiform encephalopathy is another example of a prion disease.

Biotechnology also pushes the limits of the definition of life. Geneticists are capable of creating new or novel life forms that would not normally exist in nature. This ability conflicts with an organism's ability to pass along inheritable information in a manner that maintains its lineage. It also counteracts the organism's ability to adapt through evolutionary change. Biotechnologists regularly mix the genetic material of divergent organisms to produce a hybrid, such as a potato containing particular DNA components from a bacterium or an insect. Many of these organisms are incapable of survival in nature. However, some are successful and produce a lineage of organisms that take on unusual and sometimes undesirable roles in the environment.

Scientists now have the ability to manufacture the first life form using chemical synthesis techniques. This violates the principle that living things have an evolutionary origin from a single primordial life form. In 2003, Dr. Craig Venter of the J. Craig Venter Institute in Rockville, Maryland, announced that his laboratory created an artificial virus called a bacteriophage. Bacteriophages are common viruses found in nature. They invade the cells of bacteria. Venter was able to carry out this achievement in just two weeks and showed that a simple organism can be manufactured in the laboratory using biotechnology methods. However, he cautioned that the creation of complex artificial life forms such as humans or animals is not possible with the technology of 2003.

Venter's feat, as with the accomplishments of other biotechnologists, blurs the lines between the roles of a scientist and an engineer. Hungarian physicist and aeronautics engineer Theodore von Kármán (1881–1963) distinguished a scientist from an engineer in his quote, "A scientist discovers that which exists. An engineer creates that which never was." Traditional biologists discover the characteristics of living organisms in order to better understand the principles governing nature. Much of this information is customarily used for the improvement of human life. Biologists who work in biotechnology are more like engineers as they create life forms and technologies that never existed. Biotechnology innovations led to the development of many artificial living systems that carry out adaption to the environment, evolutionary adaptation, homeostasis, metabolism, and self-replication for a variety of commercial and medical applications.

Modern biology is conducted within the framework of a paradigm centered on bioenergetics, cell doctrine, and evolution. A paradigm is a philosophy of human thought. It is essentially a predominant set of rules and regulations that establishes or defines boundaries for perceiving the world. Bioenergetics refers to the chemistry and physics principles that govern the chemical reactions taking place in living organisms. It helps distinguish between an organism and an inanimate object such as a computer. The principles of bioenergetics also help biologists understand the differences between a living and a dead organism. Cell doctrine is the theory that cells are the fundamental functional and structural constituents of all living organisms. It was proposed in 1838 by biologists Matthias Schleiden and Theodor Schwann. Evolution as proposed by Charles Darwin in 1859 is all the processes that enable populations of organisms to adapt to environmental changes from one generation to the next over a period of time.

These three principles are permanent theories of the science paradigm. However, the main ideas of these principles are not unalterable. Scientists refine these theories to more accurate representations of nature with each new discovery and innovation. But, these refinements are not always done readily. Physicist and philosopher Thomas Kuhn (1922–1996) criticized the way scientists hold on to certain outdated ideas within the paradigm of science. In his book *The Structure of Scientific Revolutions* written in 1962, Kuhn recognized that the decision to reject an existing explanation is always simultaneous with the decision to accept another. This judgment requires convincing evidence that involves the rational comparison of both ideas. The scientific community is quick to criticize biotechnology discoveries that shake the foundations of the science paradigm. Biotechnology does not suffer in its progress from this scrutiny. It improves the science of biotechnology by forcing scientists to provide credible evidence before challenging a theory that alters the science paradigm.

Bioenergetics

Bioenergetics includes the different types of chemical reactions carried out by an organism for it to maintain its characteristic life processes. All living organisms must have access to a series of chemical reactions that biologists call metabolism. Metabolism is defined as the sum of the chemical reactions that take place in living organisms. Simple organisms such as prions and viroids lack their own metabolism. As a result they rely on the metabolism of a host organism to carry out their living properties. Metabolism can be subdivided into two separate sets of chemical reactions: anabolism and catabolism. Anabolism includes chemical reactions that synthesize molecules for an organism. Catabolism represents the chemical reactions responsible for the breakdown of molecules. The term biotransformation is generally used to describe the chemical modifications carried out by living organisms. This is in contrast to the abiotic chemical reactions carried out by nonliving things. The term abiotic refers to inanimate features of nature such as climate, rocks, and water. Machines and technology are artificial abiotic things.

Almost all of the metabolic chemical reactions of organisms are carried out by special functional proteins called enzymes. Enzymes facilitate the progress of chemical reactions that would not normally occur in a manner that is favorable to life. They carry out chemical reactions by converting a molecule called a substrate into another molecule called the product. Certain enzymes break down biological molecules in a reaction called hydrolysis. Hydrolysis means to break (lysis) with water (hydro). Water is required for the hydrolysis reaction to occur. The products of these enzymes are simple molecules that serve as cell fuel or as raw materials. Another group of enzymes are involved in building or synthesizing new molecules. These enzymes are called synthetases and build complex molecules called polymers. Polymers are used to build cell structure and form storage molecules.

A special group of enzymes modify molecules by processes called oxidization and reduction. An oxidized molecule loses an electron or a hydrogen ion from its molecular structure. An oxygen atom can also be added to a molecule as it is being oxidized. Reduced molecules gain an electron or a hydrogen ion to its structure. An ion is an element or molecule having an electrical charge. Individual elements including many metals can be oxidized or reduced thereby giving them an extra positive or negative charge to the atom. All of these processes provide a direction for an organism's metabolism. Biotechnology researchers can exploit these enzymes as a way of producing electricity from metabolic processes. A team of scientists and engineers at Rice University and the University of Southern California are creating bacteria-powered fuel cells that could power small electronic devices. These devices make use of enzymes that pass electrons to metals to produce an electrical potential.

Anabolic reactions are usually carried out to help an organism maintain its chemical structure and accumulate a surplus of molecules that can be stored for later use. Biotechnology makes use of the diverse anabolic reactions that produce carbohydrates, lipids, nucleic acids, and proteins. Many of the anabolic activities that are normally carried out in a cell can be performed outside an organism using a biotechnology method called artificial metabolism. Scientists have learned to modify enzymes and metabolic pathways to synthesize novel types of molecules that are not created in living organisms. This is an excellent strategy for producing commercial chemicals with specific characteristics. The modification of metabolic pathways to synthesize molecules is called metabolic engineering.

Some examples of metabolic engineering include an underwater glue being developed by modifying certain anabolic pathways of oysters that produce a substance used to attach their shells to rocks. Biotechnology laboratories that work with bacillus bacterial are metabolically engineering the bacteria to secrete polymers that can be used as biodegradable plastics. The anabolic pathway of most interest in biotechnology is protein synthesis. Protein synthesis is the process in which amino acids are connected to each other by peptide linkages in a specific order to produce proteins. A cell's genetic material contains the code for building proteins. Scientists working in biotechnology laboratories have the skills to control protein synthesis by modifying an organism's DNA. They can also alter the genetic code for enabling a cell to produce novel types of proteins. Enzymes are probably the most commonly synthesized proteins produced using metabolic engineering.

Numerous enzymes are being used for commercial purposes. Cellulases are used to soften cotton materials in the textile industry. They break down cellulose fibers that give cotton materials a rough feel. Cellulases and related enzymes are also used to prefade clothing by removing excess textile dyes that are attached to the fabric. Amylases are also used in the textile industry to digest the starch added to blue jean fabric. Starch is added to the denim fabric to help with the cutting and shaping of blue jeans. Invertase is used in the food industry to convert glucose into fructose. Many dieticians believe that fructose is a healthier source of energy in foods and is safe for people suffering from diabetes. Proteases are used for a variety of purposes including contact lens cleaner, stain removers in laundry detergent, antifoaming agents for pools, and meat tenderizers. These enzymes digest proteins by converting them into amino acids. Lactase is used to break down the sugar lactose in cheeses and milks. This enzyme makes dairy products edible for people with lactose intolerance. The biotechnology industry makes use of thousands of enzymes in commercial, medical, and research applications.

The series of catabolic chemical reactions of primary importance in biotechnology is cellular respiration. Cellular respiration is the extraction of energy for a cell using the chemical breakdown of stored food molecules. Many cells carry out a type of cellular respiration called aerobic respiration. This type of respiration involves the use of oxygen to release energy from food molecules. It is a sequence of steps that take place within the cell. Another type of cellular respiration is called anaerobic respiration or glycolysis. Glycolysis is defined as the oxidation of molecules to produce energy in the absence of oxygen. The oxidation reaction performed in aerobic respiration combines oxygen with food molecules to cause a chemical change in which atoms lose electrons.

Anaerobic respiration in many organisms is linked to another metabolic pathway called fermentation. Fermentation is an energy-capturing process that produces a variety of molecules that are commonly used as commercial and medical products for biotechnology.

Biotechnology companies take advantage of the fermentation of bacteria, fungi, and certain animal cells for the production of commercial chemicals. Ancient people used fermentation of yeast to produce alcoholic beverages such as beer, mead, wine, and sake many thousands of years ago. These were some of the first biotechnology fermentation products. Early cultures used the fermentation of bacteria to produce a substance called lactic acid that provides the sour taste for cheese, ice cream, pasteurized milk, and yogurt. The fermentation products of filamentous fungi are used for the preparation of hoisin sauce, kimchi, poi, and soy sauce. Vinegar, or acetic acid, is another fermentation product produced by fungi including yeast. Commercial fermentation operations are used to produce a variety of chemicals including acetate used in adhesives and plastics, butyrate used for medications, glycol used in antifreeze, and propionate used for animal feeds. Tens of thousands of types of fermentation products produced by biotechnology processes find their way into everyday life.

The term fermentation is often incorrectly used to refer to any biotechnology process that takes advantage of metabolic engineering. However, true fermentation involves growing the cells in the absence of oxygen. Cells grown in the lack of oxygen modify their metabolism to reduce the production of certain cell products in favor of others. This effort conserves energy for the cells and reduces the chances of the cell backing up its metabolic pathways. Some commercial biotechnology chemicals that are produced by aerobic respiration, but are erroneously called fermentation products, are amino acids, antibacterial agents, antibodies, carbohydrates, enzymes, hormones, lipids, organic antifungal agents, peptides, pharmaceuticals, and vitamins.

Cell Doctrine

Cell doctrine, which is also called cell theory, is currently the accepted way of describing the fundamental structure that an organism needs to carry out life processes. Biotechnology views the cell as if it were a machine that can be controlled and modified to carry out specific tasks. Metabolic engineering requires knowledge of the cell components that carry out the various aspects of a metabolic pathway. Cell structures can be individually engineered to modify an organism's metabolism. Moreover, cell components can be added or subtracted to change the metabolic characteristics of a cell. Scientists have reached the point of creating artificial cells. In 2003, a team of researchers working with the National Aeronautics and Space Administration (NASA) developed an artificial cell that can carry out the metabolic functions of a red blood

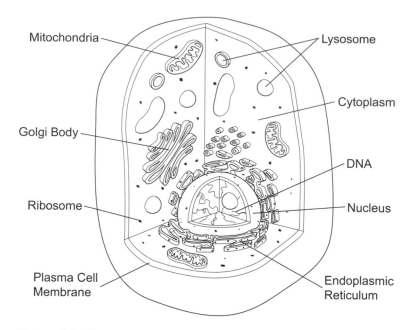

Figure 2.3 The complex cells such as those found in humans are composed of a variety of structures that contribute to the structure and functions of the body. Scientists who carry out biotechnology are particularly interested in the DNA located in the nucleus. (*Jeff Dixon*)

cell. This cell is important as a safe blood substitute for animals and humans. Other types of simple artificial cells called liposomes are a promising area in biotechnology. They are finding a variety of uses as artificial tissues and biological robots.

Scientists partition cells into three components: cell membrane, cytoplasm, and genomic material. The cell membrane is the lipid and protein covering that surrounds the cell and is involved in transport of material into and out of the cell. Cytoplasm makes up the contents within the cell membrane. The term genomic material refers to the complete heritable material, which is usually DNA, passed down from one generation to the next. Cells vary greatly in the complexity and use of these parts. As mentioned earlier, biologists accept that not all living organisms have a cell as the basic unit of structure. Viruses, viriods, and prions are disease-causing agents that are not cellular and have a simple chemistry as a basic unit of structure. Viruses are infectious agents composed of just a genome in a protein coat. Viroids are merely short pieces of RNA. Prions are the most puzzling organisms because they are no more than a piece of protein resembling abnormal proteins found in other organisms.

Microbes include any of a diverse group of simple organisms that must be viewed with a microscope. The term microbe is another word for microorganism. Bacteria, fungi, prions, protista, viroids, and viruses are all categorized as microorganisms. Bacteria are the most common microorganisms used in biotechnology. A majority of bacteria help the body and do not cause disease. They are defined as single-celled organisms that have a very simple cell structure and a circular genome composed of DNA. The cells of bacteria are called prokaryotic cells. Prokaryotes, or microorganisms that have prokaryotic cells, are characterized by genomic material that is located in the cytoplasm in a region of the cell called the nucleoid.

Another prokaryote characteristic is that their cytoplasm has no specialized compartments. Prokaryotic cells are usually a thousand times smaller than those found in the human body. This makes it very difficult to view the fine details of bacteria. Many bacteria possess swimming appendages called flagella. Almost all bacteria have a structure called the cell wall covering the cell membrane. The characteristics of the cell wall determine the way bacteria carry out certain types of metabolism. Bacteria produce a wide array of secretions that have important biotechnology applications. Many of the secretions are digestive enzymes or compounds called metabolites that are used in a variety of commercial and medical purposes. Antibiotics were the first bacterial metabolites produced as a medicine. They are still a major focus of many biotechnology industries.

Fungi are another group of cellular organisms used in many biotechnology processes. They are defined as a diverse group of organisms ranging in form from a single cell such as yeast to a mass of branched elongated and stringy cells found in filamentous fungi. Single-celled fungi are usually called yeast. Filamentous fungi are characterized by a mass of stringy cells. The cells of fungi are categorized as eukaryotic cells. Eukaryotic cells have their DNA enclosed in a structure called the nucleus. This makes the DNA nearly inaccessible to many of the biotechnology techniques involving genetic manipulation. As a result special procedures are used to reach the DNA in the nucleus. Eukaryotic cells differ from prokaryotic cells because their cytoplasm is compartmentalized into specialized functional units called organelles. Most fungi produce specialized reproductive structures, such as mushrooms. The striking colors noted on spoiled foods are mostly the reproductive structures of fungi. Several types of fungi are commonly found living harmlessly on the skin and in the digestive system. However, under certain conditions these fungi can cause mild to fatal diseases.

Protista, another group of eukaryotes, are primarily associated with diseases such as malaria and sleeping sickness. They comprise both animal-like and plant-like unicellular organisms that have limited uses in biotechnology. Algae are the most commonly exploited protista in biotechnology applications. A variety of special carbohydrates and vitamins are extracted from algae growing in large cultures. Many of these carbohydrates are used as food-thickening agents and are also used in gelatin-like desserts. The absorbent materials of disposable diapers and headbands use algal carbohydrate products. Algae are also grown in biotechnology operations called aquaculture for use in human foods and animal feed. Aquaculture is defined as the business and science of cultivating freshwater and marine algae, fish, plants, and shellfish under controlled conditions in an indoor facility or outside in penned-off areas.

Unlike microorganisms, all animals and plants are composed of eukaryotic cells. Organisms have a diversity of cell types that vary based on their cell components, functions, and shape. This means that there is no typical eukaryotic cell. Each type of cell is said to be differentiated. Differentiation means that the cell's DNA is directed to carry out a particular set of functions that contribute to the organism's homeostasis. This diversity is achieved by the way the cell's genetic material adapts the cell membrane and organelles to carry out specialized jobs. The cell membrane is a continuous double layer of phospholipids stabilized by cholesterol molecules. It encloses the contents of the cell and simultaneously acts as a two-way selectively permeable transport system. Floating around the membrane, embedded in the lipid layer, are cell membrane proteins. This "ocean" of proteins is called the Fluid Mosaic Model. Fluid describes the motion of the proteins in the membrane. Mosaic refers to the fact that the membrane is composed of a variety of proteins that appear arranged in a patchwork pattern.

It is within the cytoplasm that the cell carries out the metabolic reactions for homeostasis. Cytoplasm is divided into the cytosol and the organelles. Cytosol is a gel-like fluid composing over half of the cell's total volume. It contains thousands of enzymes that conduct a variety of cell functions mostly associated with the metabolic reactions for obtaining cell energy. Most of the chemical reactions in the cytosol are regulated by chemical information from the genomic material and the cell membrane. The organelles in the cytosol perform specialized cell functions.

Cytosol is very important as it conveys information from the cell membrane to the DNA. It helps the DNA respond to environmental signals received by the cell membrane or special proteins called receptors in the cytoplasm. Biotechnology researchers have learned how to alter the cell membrane and cytoplasm as a method of controlling the cell's DNA. It is possible to control the production of particular types of cell products by cultivating the cells in specific environmental conditions that favor one type of metabolism over another.

A group of five organelles common to most eukaryotic cells form a succession of membrane structures involved in the manufacture and movement of molecules and cell parts. These structures are collectively called the endomembrane system and include the nuclear envelope, endoplasmic reticulum, Golgi body, vesicle, and cell membrane. Components of this group transfer materials to each other through direct contact and through the use of transport vesicles. The nuclear envelope is responsible for transmitting genetic information. It also permits the inward passage of chemicals that control genetic material function. The endoplasmic reticulum or ER is an extensive network of membrane tubes derived from the nuclear membrane and connecting to the cell membrane. It is responsible for the production of the protein and lipid components of most of the cell's organelles.

A region of the ER called the rough endoplasmic reticulum (RER) usually lies closest to the nuclear membrane and is responsible for manufacturing proteins in a process called gene expression. Complex structures called ribosomes carry out this job for the RER. Ribosomes are composed of nucleic acids and proteins. Most of the proteins made in the RER are secreted from the cell. The smooth endoplasm reticulum (SER) has a variety of functions including carbohydrate and lipid production. Some cells contain a region of the ER called ergastoplasm. The ergastoplasm is a system of sack-like membrane folds in areas where the ER is continuous with the plasma membrane. This is a very important organelle for biotechnology because it is associated with cell secretions that can be used in many medical applications and as pharmaceutical compounds.

Next to the SER is a structure called the Golgi body which was named after the 19th-century Italian physician Camillo Golgi. It is also called the Golgi apparatus or Golgi complex. There can be many Golgi bodies depending on the cell's function. It is responsible for modifying, storing, and shipping certain cell products from the ER. Transport vesicles move the products from the ER to the Golgi body. Cells that specialize in producing secretions usually have a large number of Golgi bodies.

The Golgi body also produces vesicles that carry out specific chemical reactions. A lack of some of these vesicles is the basis of many human diseases. Another specialized vesicle called the lysosome contains enzymes capable of digesting the cell from inside out. These organelles recycle cell components and can be activated to cause cell death if needed. Cells can program their own death using a strategy called programmed cell death or apoptosis which is studied in many biotechnology laboratories. Cancer researchers are currently investigating biotechnology strategies that selectively cause cancer cells to undergo apoptosis.

Vacuoles are related to vesicles except that they are produced by the cell membrane. They are mostly for storing materials produced in the cell or taken in the cell membrane by a process called endocytosis. The vacuoles of plants are very valuable in plant biotechnology research. Plants can be genetically modified or metabolically engineered to store a variety of cell products in vacuoles. Scientists prefer to use plant cells for manufacturing pharmaceutical compounds because plants will not unintentionally carry diseases that are harmful to animals and humans. Pharmaceutical products made in animal cells have been known to contaminate the drugs with toxic chemicals, prions, and viroids that are difficult to detect and remove. A group of scientists at IPK-Gatersleben, a genebank in Germany, has also focused on producing spider silk in plants and is working to express complete spider silk fibers. The silk is deposited in vacuoles and is then easily harvested from the cells to be used as high-strength textiles such as those used in bulletproof vests. Plants can also have their vacuoles metabolically engineered to make the plants tolerant to drought and salt water. Other plants have vacuoles modified to store radioactive materials that are taken up from the soil and water.

Another group of organelles are called endosymbionts. An endosymbiont is a prokaryotic organism that lives within the cells of another organism. It forms an important relationship called endosymbiosis formed from the Greek words *endo* meaning inner and *biosis* meaning living. The endosymbionts usually originate from the egg's cytoplasm. This means that most organisms get these organelles from the female parent. Endosymbiont organelles contain genetic material and work in cooperation with the cell's genome. The health of a cell is monitored by information transmitted between the endosymbionts and the cell's nucleus. For example, irreparable damage to a cell triggers a response in which the nucleus and endosymbionts work together to destroy the cell. Endosymbionts are important in biotechnology because they can be cultured outside the cell for producing a variety of medically important chemicals. In addition, endosymbionts can be genetically and metabolically engineered.

The two major endosymbiont organelles of interest to biotechnology are mitochondria and chloroplasts. Mitochondria carry out aerobic respiration for a cell. They take in oxygen and simple molecules from the cell to produce much of the energy needed for cell function. Mitochondria give off carbon dioxide and water as waste products. Eukaryotic cells can have hundreds of mitochondria. They help determine the metabolic rate and energy needs of an organism's cells. Mitochondria will take on different appearances and jobs depending on the type of cell in which they are located. Mitochondria have been genetically engineered to help cells better carry out energy production. Some researchers have metabolically engineered mitochondria to produce electricity.

Chloroplasts are plant endosymbionts that carry out the metabolic process called photosynthesis. Photosynthesis is an anabolic process in which chloroplasts, with the aid of a chemical called chlorophyll, convert carbon dioxide, water, and inorganic substances into oxygen and organic compounds needed for plant structure and function. It gets its name from the fact that it uses sunlight, hence the prefix *photo* for energy. The "synthesis" part of the term refers to the fact that the energy obtained from sunlight is used to build molecules such as carbohydrates, lipids, nucleic acids, and proteins. Algae also have chloroplasts that vary greatly from those in plants. Chloroplasts amongst different algae also vary in the way they carry out photosynthesis. Many types of new crop plants are produced by genetically or metabolically engineering the plant's chloroplasts. Chloroplasts have also been altered to produce biotechnology products and electricity.

The cytoskeleton is an endosymbiont organelle that is a meshwork of protein filaments in the cytoplasm that gives the cell shape and capacity for movement. Additionally, it coordinates the function of centrioles, cilia, and flagella. Centrioles assist the cell with reproduction. Another component of the cytoskeleton found in certain types of cells is the cilia. Cilia are hair-like processes on the cell membrane and are capable of rhythmic motion. This motion helps to move body fluids on the surface of the cell including the lining of mucus inside the respiratory system. Flagella are independent endosymbiont organelles that work closely with the cytoskeleton. They are only found in protista and in the sperm of many organisms. They give sperm the ability to swim in the environment and in body fluids. These organelles are mostly of interest to biotechnology researchers who produce pharmaceutical compounds designed to alter the function of the cytoskeleton. One type of male contraception uses a biotechnology compound that alters the function of flagella and thus prevents the sperm from swimming.

The nucleus is sometimes called the "brain of the cell." This interpretation is not quite accurate. The genetic material housed within the nucleus is more like an instruction manual than a brain. With a few exceptions, every cell of the body contains a nucleus carrying an identical set of genomic information. The main role of the nucleus in the cell is genetic expression. This is a process by which the genetic material's coded information is used to produce cell structures and carry out cell physiology. Protein synthesis or gene expression is the characteristic activity of genetic function that originates in the nucleus. It is defined as the process by which cells build amino acids into proteins according to genetic information contained within that cell's genome. Many proteins build the structural features of a person while hundreds of enzymes give humans their metabolic characteristics.

The genetic code is the basis of DNA information. DNA information is organized in information units called genes. A gene can be defined in many ways. It is usually interpreted as a functional unit of heredity consisting of a segment of DNA located in a specific site of the genome. Each organism has a characteristic number and complexity of chromosomes. Chromosomes are thread-like collections of genes and other DNA in the nucleus of a cell. The term chromatin is used to describe chromosomes that are being used to run differentiated cells. Genetic engineers are skilled at altering the gene information coded in the DNA. In addition, certain biotechnology applications involve making artificial genes by synthesizing strands of DNA that are then placed into cells.

There are three major types of code in DNA programming: regulatory DNA, structural DNA, and junk DNA. Regulatory DNA is composed of chromosome segments and whole genes that function to regulate the expression of other genes. Structural genes carry the code for structural polypeptides and enzymes that build other structural components of a cell. Junk DNA is a common type of genetic information that either has no definitive role or helps reducing the effects of environmental factors that damage DNA. Certain types of junk DNA provide flexibility in the genetic code. This type of DNA is regularly exploited in biotechnology investigations involving metabolic engineering.

Gene expression is composed of two stages. The first stage, which takes place in the nucleus, is called transcription. This stage copies a particular sequence of DNA to fulfill a cell's needs. Every three sequential nucleotide bases in the DNA molecule form a "code" to match a specific amino acid, and thus each "trio," or triplet, of bases is known as a codon. For example, the DNA code ACC programs for the UGG codon. This codon is the information for the amino acid tryptophan. The order of codons in a section of DNA determines the amino acid sequence in a protein. The copied segment of DNA derived through transcription is a nucleic acid known as messenger RNA, or mRNA. The next stage of gene expression is called translation. Translation takes place on ribosomes located either in the cytoplasm or the ER. The process of translation uses mRNA to direct the synthesis of proteins from amino acids.

Gene expression begins when information from the environment or from within the cell communicates the need for a gene product. Information from the environment is either detected by the cell membrane or communication proteins inside the cytoplasm. Regulatory proteins or transcription factors are usually produced in response. These proteins locate pieces of DNA called gene regulatory networks (GRNs) that are the on and off switches of genes. The double helix of the DNA is unraveled or unzipped to expose the genetic code, which in humans is located on only one strand of the DNA, the sense strand. Antisense refers to the strand that does not code for gene information. This type of RNA carries the complementary sequence of the sense strand and serves as a blueprint for reducing genetic errors when DNA is somehow damaged. Certain biotechnology applications exploit the information of the antisense strand to produce compounds that alter cell function.

Once the sense strand is exposed, numerous types of proteins help carry out transcription. Transcription, as indicated above, involves the synthesis of mRNA using DNA as the blueprint. Transcribed mRNA is really in a form called pre-mRNA. Pre-mRNA contains alternating segments of genetic information called introns and exons. Introns are noncoding sequences of junk DNA interspersed among the protein-coding sequences in a gene. They are removed from the mRNA sequence before translation occurs. Various diseases can result from errors in this deletion process. Exons are the protein-coding DNA segments of a gene which remain after the removal of introns. They are joined together while still in the nucleus to form the resulting mRNA, which is then sent out across nuclear envelope to ribosomes either in the RER or in the cytoplasm. Introns are valuable in genetic modification procedures that alter the intron and exon sequences as a means of producing novel genes that give organisms new commercially and medically important characteristics.

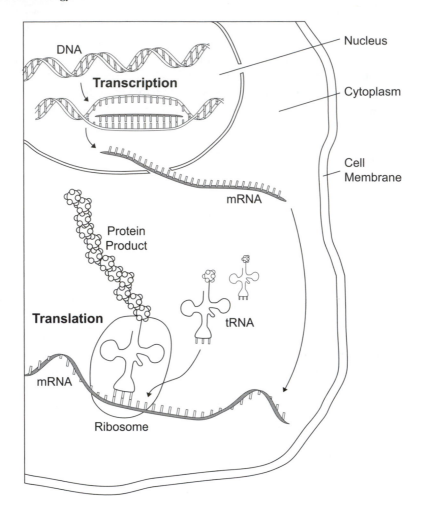

Figure 2.4 DNA programs for the production of proteins in cells. Certain biotechnology procedures modify the amounts or types of proteins reproduced by cell. This can be done as a form of therapy or as a way to get cells to make certain valuable products. (*Jeff Dixon*)

The mRNA molecule then enters the translation stage. In this stage, the mRNA binds with a ribosome and a host of molecules called transfer RNA or tRNA. Transfer RNA has structures with three nucleotide sequences that are complementary to the codon sequences of mRNA. These sequences are called anticodons. Their job is to bond with specific amino acids and transfer them to the respective codons on the mRNA.

This "matching" of codon and anticodon occurs on the ribosomes and allows the protein's amino acid sequence to be assembled according to the genetic code of the DNA. Many proteins can be made this way using one mRNA molecule. The resultant proteins are then modified and carried to particular regions of the cell. Scientists recently discovered that mRNA is also transcribed from the noncoding strand of DNA. This antisense mRNA is believed to regulate the rate of translation in a cell.

Proteins meant for secretion are synthesized in the RER and sent to the Golgi body for packaging and transport out of the cell. Bacteria carry out protein synthesis in a manner similar to eukaryotes. The ability to modify the whole sequence of gene expression is the basis of many types of biotechnology procedures. One group of techniques called RNA interference or RNAi methodologies modifies the function of mRNA as an attempt to regulate specific gene functions without altering the DNA or disrupting the function of other genes. RNAi is also being used as a biotechnology strategy for determining gene function. Scientists have also been using RNAi to knock out the function of genes that may be associated with particular genetic disorders of animals, humans, and plants. Scientists have learned that modified mRNA is blocked from translation or degraded by an enzyme called dicer that usually protects the cell from certain viruses called double-stranded RNA viruses.

Evolution

Cell doctrine overlaps with the theory that life comes from preexisting life and is subject to adaptive changes from one generation to the next. Cells live in one of three stages of existence. Active cells are said to be vegetative or differentiated meaning that they carry out a particular task to stay alive. The cell making up a single-celled organism must be able to carry out all the metabolic tasks needed to perpetuate its existence in a particular environment. Multicellular organisms are generally composed of cells that carry out a specific set of tasks that contribute to the organism's survival. Differentiated cells are sometimes said to be a G_0 stage. Some cells are dormant meaning that they either are not carrying out metabolism or are not performing a function for the body. Many bacteria, fungi, and protista produce dormant cells called spores. Spores permit the cells to evade damaging environmental changes that could dehydrate, freeze, or overheat active cells. The dormant cells of multicellular organisms can exist as germ or stem cells. These are cells that can give rise to other cells or can differentiate into a particular type of cell.

A stem cell is defined as an undifferentiated cell that can make similar copies of itself indefinitely and can become specialized for various functions in an organism. Stem cells have various abilities of differentiating. Unipotential stem cells are capable of differentiating into one particular type of cell. A limited number of cell types representing a particular category of cell in an organism can form from multipotential stem cells. Pluripotential stem cells can develop a large variety of cells and can even form tissues and organs. Totipotential stem cells are capable of forming a whole organism, which makes these cells favorable for cloning and stem cell research. The differentiation of cells requires many signals from the environment and other cells in order to start the development process and progress into a particular cell. Researchers are currently finding ways of directing cells to carry out many functions in cultures and in living organisms. Stem cell research is a rapidly growing area of biotechnology.

Cells that are not differentiated or dormant are usually taking on the duty of replication. They carry out a sequence of stages called the cell cycle. During the cell cycle, the cell passes through one cell division and the next. Cell replication takes place when signals called mitogens initiate the process of cell division. Asexual division is a type of reproduction in which two new cells develop from a single cell. Bacteria undergo a process called binary fission to carry out sexual production. It is a simple process that replicates the genome and cell contents to make two identical bacterial cells. Eukaryotic cells undergo a type of asexual reproduction called mitosis. It is a complicated series of events.

Some cells undergo sexual division. Sexual division carries out a type of chromosome replication and cell divisions that result in the formation of cells called gametes. Many eukaryotes have two copies of DNA and have a genetic condition called diploid genetics. The diploid condition is defined as cells having a full set of genetic material consisting of paired chromosomes. Each pair of each set or homologous pair of chromosomes represents a parental set passed along by sexual reproduction. Most animal and plant cells are diploid. Gametes or sex cells are haploid cells and contain half of the chromosomes of a diploid cell. Scientists who conduct genetic manipulation must take into account whether a cell is diploid or haploid. Each copy of a gene on the homologous chromosomes in a diploid cell must usually be altered to ensure the characteristics are changed. Haploid cells only have one set of DNA, which means that any genetic change induced in the cell is expressed with no competition from a related gene.

The cell cycle of eukaryotes can be divided into two main stages: the interphase which prepares the cell for replication and the M phase where nuclear and cytoplasmic division occurs. Interphase is divided into several steps called the G_1 phase (gap 1), the S phase (synthesis) in which DNA replication occurs, and the G_2 phase (gap 2). The M phase is a sequence of events that includes prophase, metaphase, anaphase, and telophase. There are two types of M phases in the cell cycles: asexual division (mitosis) or sexual division (meiosis). Interphase is a common stage of the asexual and sexual cell cycles. It is during interphase that the DNA is replicated in the nucleus in preparation for division stages. Eukaryotic cells can spend up to 20 hours in interphase.

The cell produces the components of the cytoplasm and the various enzymes needed for cell division while in the G_1 stage. Cells require many nutrients during the G_1 phase and this becomes an important factor when keeping cells alive and healthy in cell cultures used for biotechnology applications. During the S stage the cell doubles its DNA content as an outcome of chromosome replication. This doubled DNA is called the chromosome. Each half of the doubled chromosome is called a chromatid and is an exact copy of the other. DNA is inactive at this point and is very prone to alterations in the genetic code as it is replicating. These changes in the genetic code are called mutations and are of interest to geneticists who work in biotechnology. Mutations are essential for providing genetic variety that organisms need to adapt to environmental changes from one generation to the next. They are also important because mutations can impart novel traits to an organism that may have important commercial or medical applications. The G_2 phase carries out final preparations for the cell division phases.

The G_2 phase heralds the end of interphase and the beginning of the M phase which starts out with prophase. During prophase the doubled chromatids are attached to one another at a region called the centromere. This makes up the structure called the chromosome. The chromosome now contracts into a compact tightly coiled structure called heterochromatin. Biotechnology researchers have learned that twisting certain sections of DNA in heterochromatin can shut down the expression of a particular trait. This is then followed by the breakdown of nuclear envelope that releases the chromosomes into the cytoplasm. Proteins called spindle fibers begin to form and attach to the centrioles. The centrioles then start to separate and move apart in opposite directions in preparation for dividing cell components into opposing regions of the cell.

Metaphase follows prophase. In metaphase, the chromosomes are pulled into a flat line midway between the two centrioles that are now at opposite ends, or poles, of the cell. This midline is called the equatorial plane and represents the region where the whole cell will divide into two. The chromatids now attach the spindle fibers to the centromeres. Mitochondria and chloroplasts are also attached to spindle fibers. Plants seem to lack centrioles. Thus, their spindle formation is under the control of another cell component. Spindle formation is related to other cytoskeleton functions that contribute to cell function. This feature of the cell is of special interest when cells are metabolically engineered for biotechnology applications. Metaphase provides geneticists an opportunity to count and identify the different chromosomes of an organism. They are highly visible at this stage and lined-up for easy viewing.

Anaphase starts to progress at the end of metaphase. During anaphase the two chromatids of each chromosome begin to separate, moving to opposite ends of the cell. They are pulled along the spindle fibers by the centromeres. Genetic errors are likely to occur during anaphase resulting in too many or too few chromosomes in the resultant cell. Several genetic disorders of animals and humans result from this condition and is of interest to biotechnology researchers who study and try to correct genetic abnormalities. Anaphase is immediately followed by telophase. In telophase, a new nuclear envelope forms around the separated DNA at each end of the cell. Now the spindle fibers disappear as the chromosomes uncoil.

The separation of the DNA into different nuclei is called karyokinesis. A result of this process could be described as a double-nucleated cell. Cells with two or more times the DNA of a usual cell have important commercial applications in biotechnology. They can be induced to produce large amounts of a gene product that is collected and purified as a biotechnology product. In order to actually produce two separate cells, a process called cytokinesis has to occur. Cytokinesis is the division of the cytoplasm after karyokinesis has occurred. Cells having completed the M phase can either reenter dormancy, differentiate, or undergo another round of division. Stem cell researchers need to have stringent culture conditions that control the cell's fate after it completes a cell cycle. A researcher needs to know when and how to guide a culture of cells into division or differentiation to work successfully with any cell culture.

The term meiosis, or reduction division, was derived from the Greek word "decrease." Early biologists, viewing what they thought were cells undergoing mitosis, noticed a strange sequence in which the amount of

DNA halved after two cell divisions. This type of division only occurred in gamete-producing cells. Therefore, it was hypothesized that the cell division being viewed was a method of decreasing the DNA content for the formation of gametes. Meiosis starts out with an interphase that leads into two stages of nuclear division. The stages are called meiosis I and meiosis II. Special mitogens turn on genes that direct the cell to undergo meiosis. Meiosis I is divided into four stages: prophase I, metaphase I, anaphase I, and telophase I.

Prophase I is almost identical to the prophase stage of mitosis. The main difference is that during prophase I the chromosomes arrange into homologous pairs. Homologous chromosome pairs have the same lengths, the same centromere positions, and in most cases, the same number of genes arranged in similar linear order. It is possible at this time for the maternal and paternal chromosomes to swap segments of DNA in a process called crossing over. During prophase, homologous chromosomes are paired together and situated close to each other. Certain segments along the chromosomes make contact with the other homologous pair. This point of contact is called the chiasmata and can allow the exchange of genetic information between chromosomes. This further increases genetic variation needed for survival from one generation to the next. Biotechnology researchers depend on crossing over as a way of obtaining genetic variety in organism breed for biotechnology applications.

In metaphase I the centrioles attach spindles to only one set of the chromosomes. The spindle fiber of one pole is attached to the maternal chromosome while the spindle at the other pole attaches to the paternal chromosome. Metaphase I lines up the homologous chromosomes to ready them for separation during anaphase I. Anaphase I then separates the maternal and paternal pairs to opposite poles of the cell. At the end of telophase I, each cell has half the number of chromosomes but each chromosome consists of a pair of chromatids. Meiosis II then jumps into metaphase II and anaphase II, which line up and separate the chromatids. Metaphase II is essentially the same as mitosis in that chromatids of each chromosome are being separated. By the end of telophase II, four gametes are formed.

Biologists in the 19th century could only speculate about the roles of mitosis and meiosis in perpetuating an organism. Little was known about genetics or the chemistry of DNA. It took a radical view of nature to prompt the scientific community to investigate the mechanisms of inheritance long practiced by selective breeding of animals and plants on farms throughout the world. Selective breeding is defined as breeding

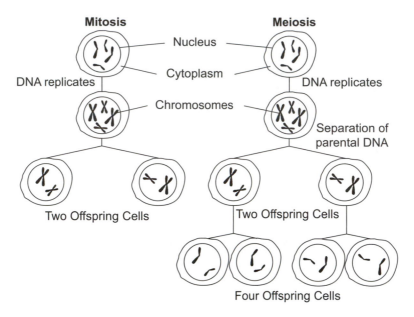

Figure 2.5 Mitosis is a means by which a cell replicates to produce two identical cells. Cells use meiosis to produce gametes such as eggs and sperm. (*Jeff Dixon*)

an organism that has a desirable trait with another so that the particular trait is passed to the next generation. In 1959, the naturalist Charles Darwin published a book titled *On the Origin of Species by Means of Natural Selection, or the Preservation of Favoured Races in the Struggle for Life.*

In the preface of this book, Darwin made a statement that was a growing sentiment in the scientific community, "I will here give a brief sketch of the progress of opinion on the Origin of Species. Until recently the great majority of naturalists believed that species were immutable productions, and had been separately created. This view has been ably maintained by many authors. Some few naturalists, on the other hand, have believed that species undergo modification, and that the existing forms of life are the descendants by true generation of pre-existing forms." In brief, Darwin took the lead in explaining a mechanism for how genetic variation arises in organisms and how new species appear to arise. His observations provided the impetus scientists needed to further investigate the chemistry and functioning of the inherited material.

Darwin reinforced the emerging idea of natural selection by reinforcing a rising opinion of naturalists with his meticulous observations of animals, fossils, and plants studied on his voyages on the *HMS Beagle* between 1828 through 1836. His most enlightening reflections took place on the Galapagos Islands located off the coast of Ecuador. The theory of natural selection contradicted public views of the creation of organisms and implied that all species originated from common ancestors through predictable processes of nature. Natural selection is considered to be the biggest factor resulting in the diversity of species and their genomes.

Darwin's principles of natural selection are categorized into four basic principles that he supported with his observations:

1. One of the prime motives for all species is to reproduce and survive, passing on the genetic information of the species from generation to generation.

2. Organisms tend to produce more offspring than the environment can support. The lack of resources to nourish these individuals places pressure on the size of the species population, and the lack of resources means increased competition and as a consequence, some organisms will not survive.

3. The fact that organisms die as a result of competition is not a random occurrence. Certain organisms are more suited to their environment and are more likely to survive. Organisms most suited, or more fit, for their environment have more chance of survival if the species falls upon hard times.

4. Organisms that are better suited to their environment exhibit desirable characteristics, which is a consequence of their genome having mutations that by accident provide the desirable characteristics. A weeding-out effect occurs that permits those organisms with the desirable traits to produce more offspring than those lacking these traits.

From these observations of living and extinct organisms, Darwin was able to conclude that organisms had evolved over time. In addition, he formulated that organisms with the most desirable characteristics for their species were favored and had outpaced the reproduction of other organisms. Thus, organisms that were better adapted to survive the variety of events that they experience in their life passed their genes on to the next generation.

It is now accepted by the scientific community that mutations provide the genetic changes that produce the variety of traits exhibited by a population of organisms. An organism's interactions with other organisms and with environmental conditions select the predominant traits of the population. Changing environmental conditions would mean that different characteristics would be favored in following generations in response to the particular environmental changes. Darwin believed that organisms had evolved characteristics that were useful for their environments. This caused the organisms to occupy an ecological position where they would be best suited to their environment and therefore have the best chance of survival. Geneticists developed the term "allele" to refer to the genetic variations of a particular trait brought about by mutation. The organisms seen on Earth today are recognized to be the result of the process of evolution over a period estimated to be 3.4 billion years.

Scientists who preceded Darwin promoted the theory of natural selection without a full understanding of the genetics behind inheritance and mutation. DNA's role as the genetic material was not full recognized until the 1960s with the discovery of gene function. The nature of the genetic changes induced by mutations was also being unraveled during that time. All of these new findings provided a chemical explanation for the mechanisms underlying natural selection. Darwin recognized that changes to the inherited material that are passed on from one generation to the next are important for natural selection. Today, scientists know that these changes are alterations of the DNA, which are called mutations. A mutation is usually defined as any heritable change in genetic material.

At its most uncomplicated expression, a mutation is a chemical transformation of one nucleic acid within an individual gene that may or may not alter its function. Severe mutations may involve the rearrangement, gain, or loss of part of a chromosome. These types of mutations can be viewed microscopically and can cause significant changes to the organism's characteristics. Large mutations such as these are called chromosomal aberrations. The simplest and most common type of mutation is the base pair or point mutation. This was one of the first mutations discovered and involves the substitution of one nucleic acid with another. Most base pair mutations do not cause any significant changes to the function of the gene. However, the debilitating disease called sickle cell anemia is due to a base pair mutation. Sickle cell anemia is the most common inherited blood disorder in African and Mediterranean people. It affects over 1 in 500 African Americans in North America. The mutation causes the normally globular hemoglobin molecule of red blood cells to clump together into rigid fibers. This in turn distorts the red blood cells and reduces their ability to carry oxygen throughout the body.

Today, most people who work in the field of biotechnology refer to base pair mutations as single nucleotide polymorphisms or SNPs (pronounced snips). The National Center for Biotechnology Information describes an SNP as the alteration of a DNA segment such that the genetic code AAGGTTA is changed to ATGGTTA. Notice that the second "A," or nucleotide called adenine, is replaced with a "T," which refers to the nucleotide called thymine. It is estimated that an SNP mutation takes place in the human population more than 1 percent of the time. The effect of an SNP on a creature depends on where the change occurs in the DNA. In humans, about 3 to 5 percent of the DNA contains the genetic code for proteins that contribute to a person's characteristics. SNPs in this type of DNA were recognized years ago as being base pair mutations because geneticists at that time mainly studied those DNA sequences.

However, it is now known that most SNPs are found outside of DNA that programs for proteins. These SNPs are important to biotechnology researchers because they are likely to affect the DNA that organizes protein production. This organizational DNA, which is usually called regulatory DNA, controls the patterns of traits expressed in an organism. A major group of regulatory DNA sequences, called homeobox or HOX regions, are responsible for organizing features such as the location of the eyes, the number and position of limbs, and the placement of organs. SNPs in these regions have major evolutionary significance and are important for understanding genetic diseases or humans and domesticated animals and plants. Significant efforts have been made by the biotechnology community to study the SNPs of humans and many animals and plants that have significance to human survival. The governments of various countries and many research universities keep online SNP databases for organisms having economic, medical, and research importance.

Other types of mutations of interest in biotechnology are insertions and deletions in which extra nucleotides are added or deleted from a sequence of DNA. The number of nucleotides can range from a few to thousands. These types of mutations can lead to genetic disorders corrected with gene therapy. In addition, these mutations are also useful for DNA analysis because they create unique patterns in an organism's DNA. Certain insertions and deletions occur in multiples of one or two and result in a frameshift mutation. These mutations usually produce life-threatening effects. The DNA alteration that results from these mutations either codes for ineffective proteins or disrupts the regulation of a trait. Frameshift mutations of HOX genes can cause the loss of major body parts and have been shown to affect the number of legs in many animals.

Frameshift mutations are very important research tools in biotechnology because they can be used to eliminate or "knock out" a particular characteristic. Biotechnology researchers grow "knock-out" animals for investigating how lack of a gene will affect other genes and the whole organism. "Knock-out" rats are used for studying and treating human genetic disorders. Many pollutants are known to cause frameshift mutations associated with cancer. A test called the Modified Ames Test is used to determine if certain chemicals used in foods, clothing, cosmetics, and medications are capable of causing these types of frameshift mutations. It is performed on bacteria grown with the liver extracts of specially bred mice. Chemicals introduced to the solution are modified by liver enzymes, in a similar manner in which this would occur in the body, and then the bacteria are tested to see if they induced a frameshift mutation. Rapid tests modeled on the Ames test have been developed using eukaryotic cell cultures and more accurately depict the cancer-causing ability of chemicals on higher organisms.

Chromosomal aberrations include deletions, duplications, inversions, nondisjunctions, and translocations. These extreme mutations are important for understanding human genetic disorders and have the potential for producing organisms with unique characteristics for biotechnology applications. Deletions involve the loss of a large section of a chromosome resulting in the loss of a number of characteristics. These mutations can be induced through genetic modification to remove undesirable traits from organisms used in agriculture and research. Duplications are turning out to be surprisingly common mutations that have important evolutionary implications. In this mutation certain genes are duplicated on the same chromosome. The hemoglobin protein used to carry oxygen in the blood is constructed from duplicated gene. These genes have much value in biotechnology because they will mutate separately and give rise to new characteristics. In addition, genetic alteration procedures that induce these mutations are performed to enhance desirable traits in an organism.

Inversion mutations are being used in biotechnology as a means of modifying the characteristics of economically important organisms. These mutations involve the rearrangement of a region of DNA on the chromosome so that its orientation is reversed with respect to the rest of the chromosome. Biotechnology researchers have learned to regulate the expression of ripening genes in fruits by inverting certain segments of DNA that control the sequence of fruit development. The Flavr Savr tomato developed by Calgene of Davis, California, in August, 1991 used an inversion mutation technology that adjusted the ripening process so that the tomato turned red and tasty without getting soft. Certain gene therapy procedures call for the insertion of inverse genes to turn off deleterious traits that result in genetic disorders.

Nondisjunctions are medically important aberrations that result from errors in mitosis and meiosis. In this type of aberration the chromosomes fail to successfully separate to opposite poles of the cell during division. This results in an uneven distribution of genes for major characteristics in the cell. Some researchers are looking at ways of inducing these mutations in cancer cells as a treatment to slow down the progression of the disease. Translocations involve the transfer of a piece of one chromosome to an unrelated chromosome. The piece of DNA that relocates another DNA segment is a regulatory gene called a transposable element or "jumping gene." Translocation is a natural event in cells that helps gene regulation. The immune system relies on this type of gene movement to assist with the identification of foreign materials in the body. However, abnormal translocations can lead to diseases such as the blood cancer leukemia. Transposable elements are used as tools for inserting segments of DNA into chromosomes in genetic modification procedures.

Mutations can affect an organism's protein expression in various ways. This is important to know when conducting genetic modification experiments intended to alter protein production. The genetic modification procedure must be done carefully so that it produces the desired effects without causing abnormalities in the organism. Mutations can be grouped according to the type of effect they have on an organism. The major categories are missense, nonsense, silents, and splice-site mutations. Missense mutations are commonly used in a variety of biotechnology applications. A missense mutation produces a genetic code change that alters a codon. This results in a different amino acid being placed into the protein of the gene in a manner that alters the characteristics or function of the protein. This type of mutation is represented by sickle cell anemia. Missense mutations are a source of new traits that can provide valuable characteristics for organisms that have agricultural and commercial value. Many biotechnology researchers search for missense mutations that produce novel proteins with potential therapeutic value.

Nonsense mutations are nucleotide changes that stop the synthesis of a protein before it is completely expressed. These mutations stop the translation of the mRNA prematurely to produce a truncated protein. Truncated proteins are not likely to function properly and may even interfere with the functions of other genes. Nonsense mutations are responsible for many genetic diseases of animals and humans. Thus, they are important to scientists who study agricultural and medical biotechnology. Silent mutations cause genetic variation by changing the nature of the protein. They are only detectable by DNA analysis. These mutations are used in biotechnology procedures that track the evolutionary origins of organisms. Some researchers are able to trace the ancestry of organisms by tracking silent mutations in the chloroplasts and mitochondria.

Splice-site mutations affect special areas of DNA found only in eukaryotic cells. These regions are called introns. An intron is defined as a section of a gene that does not contain any instructions for making a protein. Introns break up the sequence of information in a gene. They are removed from the mRNA just after transcription. This leaves behind bits and pieces of the DNA segments called exons that contain the protein coding information. The exons are bonded together to form the completed mRNA that undergoes translation to produce the protein product of the gene. This process of mRNA splicing must be carried out very accurately for the gene to carry out its proper function. Scientists are now learning that there is some variability for how introns are removed from a particular gene's mRNA. This is a means of having one gene produce two or more different proteins. Each protein in turn can help the organism carry out a different function. Mutations that alter intron removal will produce an incorrect protein that may not function properly. Biotechnology researchers value these mutations because they can impart novel commercially important characteristics in an organism.

Modern biotechnology research relies heavily on the knowledge of how genetic variation adapts organisms to environmental factors. Agricultural research conducted before the 1980s rarely took into account the ability of a crop or domesticated animal to survive changing environmental conditions such as disease or drought. So, it was common for farmers to lose almost all their crops and livestock in a season. The older methods of selective breeding made it difficult for agricultural researchers to provide agricultural organisms with desirable commercial traits as well as the characteristics that made them resilient to disease and catastrophic environmental changes. A better understanding of how mutations provided benefits to particular organisms paved the way for biotechnology techniques that could produce the "fittest" domesticated animals and plants.

Certain biotechnology laboratories purposely place organisms and cells under changing environmental conditions to bring out characteristics that are normally not seen in the wild. This is how genes for surviving freezing temperatures or drought were discovered in certain crops. Other research studies use chemicals and radiation to induce mutations. The scientists then evaluate the mutations to see if they produce valuable characteristics for an organism. A new branch of biotechnology research focuses on molecular evolution carried out in mixtures of biochemicals. Molecular evolution is the study of how molecules change and evolve as a result of specialization or selection. In some molecular evolution studies, scientists deliberately create DNA sequences that may have evolutionary value in agricultural animals and plants. One biotechnology application of molecular evolution research is a special protein that mutates so that it acts like antifreeze. One form of this protein has been discovered in fish that live in the frozen waters of the North Sea. A protein normally found in abundance in the body fluids of the fish mutates to block the formation of ice crystals in the cells and body fluids.

3

THE TOOLS OF
BIOTECHNOLOGY

INTRODUCTION

Biotechnology is an interdisciplinary science that borrows scientific instruments commonly used in chemistry, biochemistry, genetics, and physics laboratories. Very few instruments are specifically designed for biotechnology. Those that are unique to biotechnology were developed for the specific needs of particular research studies. A trip to a biotechnology laboratory would seem very much like a visit to any other science laboratory. This is also true for large facilities that produce biotechnology products. The machinery is used in many other industries. However, biotechnology instruments are focused on analyzing, manipulating, or manufacturing the chemicals that make up organisms. The major chemicals of interest in biotechnology are biological molecules called nucleic acids and proteins. Each instrument mentioned in this chapter can be found in most biotechnology industrial settings. Research laboratories are usually limited to particular equipment for research being performed.

The biotechnology tools mentioned in this chapter are integral components of the biotechnology techniques described in the next section. Most of the tools of biotechnology are used to identify and isolate many of the biological molecules making up an organism. The identification of biological molecules is called characterization. Characterization tells researchers the specific chemical makeup of a molecule. General chemical characterization techniques help scientists in identifying molecules as one of four major biological molecule categories: carbohydrates, lipids, proteins, or nucleic acids. Resolution is a term used to describe the degree of detail used to characterize molecules. For example, high-resolution characterization provides information about the specific identity of a particular type of biological molecule. Many of the tools described in the following section tell researchers whether a particular protein or sequence of nucleic acids is present in a sample. Isolation is a method of separating a particular molecule from a mixture. Researchers interested in working with a pure sample of a molecule must isolate and collect it from a mixture. Many of the tools that identify molecules also isolate that molecule from the mixture, saving the researcher time and effort.

The first biotechnology tools date back to fermentation jars used to make alcoholic beverages used by ancient people almost 7,000 years ago. Special ceramic pots designed to enhance fermentation were discovered in archeological sites throughout Asia, the Middle East, and South America. Almost 3,000 years ago the Chinese were using devices for culturing and extracting antibiotic chemicals from moldy soybean curd. A boom in scientific instruments started in Europe after the 1600s with the advent of the microscope and new apparatus for conducting chemical reactions. The harnessing of electricity to operate machines refined the instruments used in older biotechnology applications. In addition, electricity permitted scientists to develop the great variety of analytic instruments used everyday in biotechnology. By the late 1800s many of the instruments such as centrifuges and incubators seen in modern biotechnology laboratories were being developed.

Improvements in electrical circuitry, motors, and robotics further refined the types of instruments used in biotechnology. Instruments were becoming more accurate and simpler to use. The advent of computers fueled tremendous improvements in biotechnology instruments. Almost all of the instruments used in biotechnology today have a built-in computer or are linked to computers that integrate the instrument with other tools of biotechnology. Computers also make it possible to replace chart paper and older ways of collecting and recording data. This data can now be imported into other instruments or into a software that carries out various types of analyses and statistical calculations. The computer can also place the data into an electronic notebook that could be e-mailed to other scientists.

Advances in miniaturization and the creation of lightweight materials for constructing instruments are providing new directions in biotechnology instrument design. Instruments that at one time took up all of the space on a laboratory table can now fit into an area of the size of a small toaster. Portable instruments are making it possible for scientists to share and transport expensive and specialized instruments. This is particularly important in bioprocessing operations in which it is favorable to carry out instrumentation procedures at difficult locations of a facility. Miniaturization is leading to the development of microscopic instruments that can be placed into cell cultures of whole organisms for continuous monitoring. New methods of wireless communication is enhancing the ability of the instruments to transfer data. Scientists now have access to instruments that use devices similar to cell phones that can control instruments and transmit data to various computers.

THE TOOLS

Amino Acid Analyzers

Amino acids are the building blocks for proteins. There are 20 naturally occurring amino acids that commonly make up the proteins of organisms on the Earth. At least 20 others are important in biotechnology research. Many other artificial amino acids make up proteins for commerce and research. Proteins carry out their functions based on their amino acid composition. Hence, the amounts, sequence, and types of amino acids are used to characterize proteins. Amino acid analyzers are machines that provide biotechnology researchers with information about the amounts and types of amino acids making up a protein. They have many other applications in food testing, forensic evidence analysis, and pharmaceuticals development. The typical modern amino acid analyzer is a large machine run by a computer. There are various types of amino acid analyzers depending on the types of protein samples being tested. The simplest ones require that the samples are specially prepared and manually injected into a collection device. Elaborate analyzers do almost all of the work by taking raw material and preparing for the analysis with computer driven robotics.

All amino acid analyzers have one core component called the chromatography unit or column. The chromatography unit is the part that separates the different amino acids based on their individual chemical properties. Samples of proteins are broken down into amino acids and then pumped through the chromatography unit while dissolved in special solvents. Each amino acid travels through the chromatography unit at a different rate. The amino acids then pass through another part of the amino acid analyzer called the detector. The detector uses a beam of light to measure the amount of each amino acid that crosses the beam. This information is then charted on a graph called a chromatogram. The chromatogram tells the scientist the amounts of each type of amino acid found in the protein. A technique called amino acid sequencing then helps the scientist determine the order of the amino acids making up the protein. Researchers need to isolate molecules for a variety of reasons. Isolated proteins can be used as drugs. Pure segments of DNA could contain a gene that is later inserted into an organism for genetic engineering research.

Amino Acid Sequencers

The amino acid composition of a protein alone does not give the full nature of its structure. It is the sequence of amino acids in a protein that provides its major characteristics. Scientists can tell the chemistry and shape of a protein knowing its amino acid sequence. They can then use this information to calculate the approximate order of the genetic information programming for the protein. This in turn can help scientists find the location of a gene on a large segment of genetic information. Amino acid sequencers are elaborate pieces of equipment that must take apart a sample protein piece by piece in a manner that determines the arrangement of amino acids making up a protein. Amino acid sequencing was a time-intensive procedure before the technique was automated. It could take days to sequence even simple proteins. Moreover, it took a series of calculations to figure out the proper amino acid arrangement. The procedure usually had to be replicated several times to ensure accurate information. This meant more time in the laboratory doing a demanding procedure.

Automated sequencers are able to prepare the sample, break apart the protein, feed it into the analyzers, and then determine each amino acid as it is broken off the amino acid chain. It does it quickly and can carry out the procedure multiple times. The typical apparatus has a re-action area, a sample collector, a chromatography unit, and a detector linked to a computer. Traditional amino acid sequencers use a method called N-terminal sequencing. Each protein has two ends. One end is called the N-terminus and the other is called the C-terminus. The end of the protein called the N-terminus is labeled with a chemical called phenylisothiocyanate (PITC) in N-terminal sequencing. PITC serves as starting point for the disassembly of the protein. A chemical called triflu-oroacetic acid is then added to break off the PITC labeled amino acid. This is then converted into another chemical that is fed into the chro-matography unit. Each amino acid travels through the chromatography unit at a different rate. The amino acids then pass through another part of the amino acid analyzer called the detector. The detector uses a beam of light to detect whether an amino acid crossed the beam. This infor-mation is then charted on a graph called a chromatogram. The chro-matogram is a permanent record of the sequence of each type of amino acid found in the protein. It provides the best information on sections of protein no more than 50 amino acids long. So, large proteins must be chopped for study. A new technique called C-terminal sequencing was recently developed. It uses other labels and acids to sequence the protein from the opposite direction. This technique is useful on proteins that are difficult to study using the N-terminus method.

Balance

Balances are devices for accurately determining the mass of a chemical. They are not the same instrument as a bathroom scale or postage scale that measures weight and not mass. Mass measures the amount of matter making up an object. Weight is a measure of the force of atmospheric pressure and gravity on the mass of an object. Scientists do not usually use weight when measuring quantities of chemicals in the laboratory. Unlike mass, the weight of an object varies depending on the humidity, location, and temperature. So, it would be inconsistent to use weight as a method of determining chemical quantities. Many of the chemical solutions used in biotechnology are mixed using precise amounts of chemicals. These solutions must be made the same each time the procedure is carried out to ensure that the process is consistent and works properly.

Balances used in biotechnology vary greatly in size and measurement capacity. Large balances that mass the raw materials on a truck can measure thousands of kilograms of materials. Medium-sized balances measure hundreds of kilograms of chemicals or materials used in producing biotechnology products. Analytical balances were developed for measuring minute masses of chemicals and materials used in scientific research. Very sensitive analytical balances can measures masses in hundredths of a milligram. However, most small balances are used to calculate mass in grams. Analytical balances are found in every biotechnology laboratory. In addition, many types of biotechnology manufacturing equipment have built-in balances that provide the mass of materials being processed or transported during a particular procedure. Most balances are used to measure the mass of a chemical, while others are specially designed to calculate the amount of moisture in a sample. The first balances were mechanical devices that did not use electricity to operate. Almost all of the modern balances used in biotechnology require electricity to run some component of the balance. Mechanical balances were often difficult to use consistently and the accuracy of their measurements were often subject to the skills of the user.

Many analytical balances are composed of a sample pan, a beam called a fulcrum, a comparison standard, and a readout. The sample

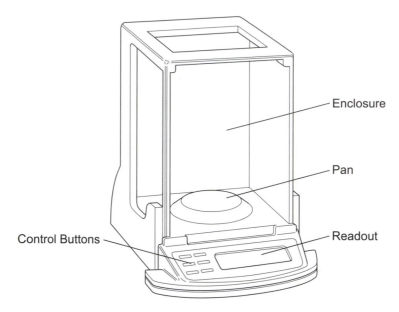

Figure 3.1 Analytical balances are precise instruments used to weigh out chemicals used in biotechnology applications. (*Jeff Dixon*)

pan is attached to one end of the fulcrum and the comparison standard is at the other end. Material being massed is placed on the sample pan. The mass of the material on the sample pan then presses on the fulcrum. Adjustments are then made to the comparison standard so that pressure is placed on the other end of the fulcrum. The function of the comparison standard is to provide a reference for the mass of the material being measured. Mass is determined when a certain amount of the comparison standard presses equally to the sample on the fulcrum. The readout shows the mass number for the fully balanced fulcrum. A growing number of balances replace the comparison standard with sensor switch having a built-in computer chip. In these balances, the sample pan presses on the fulcrum that is attached to the sensor switch. The sensor switch then compares the mass of the sample to a computer program. It then provides a digital readout of the mass based on the computer's calculation.

Chemicals and objects are usually never placed directly on the sample pan. Foil, glass, paper, or plastic weighing containers are used to hold the sample being massed. These weighing containers are usually handled with tongs or gloves to prevent chemicals and water in the fingerprints from affecting the mass reading. The mass of the container must be subtracted from the mass of the sample. The term "tare" is used to represent the mass of the weighting container. A person using the balance must first determine the mass of the tare and then reset the balance to read zero using a tare adjustment knob. They can then add the sample to the weighing container and use the new readout provided by the balance. A tare adjustment must be made every time the balance is used. It cannot be assumed that all similar weighing containers have the same mass.

All balances must be calibrated regularly to ensure they are providing the proper mass and are working consistently. Calibration is defined as the process of adjusting an instrument so that its readings are actually the values being measured. This is done by placement of special weights called calibration standards on the pan. The balance is then tested several times to see if it accurately and consistently matches the mass of the calibration standard. Adjustments to the balance can be made if the balance is not calibrated. Most modern balances have built-in calibration weights to maintain calibration. Analytical balances must be used in a draft-free location on a flat, solid bench that is free of vibrations. Balances are very sensitive to being bumped and must be used with electrical systems that do not fluctuate. Objects too heavy for the balance to mass can damage the fulcrum or the sensor switch. Some laboratories require that all measurements for one procedure are done on one particular balance to ensure any possible inconsistencies between different balances.

Bioreactor

Bioreactors are containers for culturing microbes, growing cells, or carrying out chemical reactions used in biotechnology applications. Research laboratories typically use small bioreactors that hold less than one liter of liquid. Laboratories that develop new biotechnology products use medium-sized bioreactors that can contain many liters of solution. These are commonly used in large facilities called pilot plants. Pilot testing is a series of experimental procedures that investigate whether large amounts of a particular biotechnology process can be carried out in a cost effective way. Biotechnology companies involved in the production of large volumes of materials use bioreactors that can hold thousands of liters of liquid.

Certain bioreactors are called fermentors because they carry out their job in the absence of oxygen. Some organisms carry out a type of metabolism called fermentation when oxygen is not present. Alcohol and many other biotechnology products are made using fermentation. Certain chemical reactions are inhibited by oxygen and are also conducted under fermentation conditions. Bioreactors are also referred to as bioprocessors and digesters depending on their use. Bioprocessors are used for producing a variety of chemicals from secretions produced by cultured cells. Pharmaceutical companies use bioprocessors to produce drugs such as insulin from genetically modified bacteria. Digesters contain cells or chemical mixtures that break down particular compounds and convert them to commercial products. Biofuels such as methane gas are made in digesters. Bacteria or yeast grown in special digesters break down agricultural wastes from animal or plant into the biofuels.

There is no typical type of bioreactor. Their design and function depends on the type of reaction being carried out and the type of material being produced. However, all bioreactors have several major components: atmosphere supply, collection port, control panel, media supply, mixer, and vessel. The vessel is the main component of the bioreactor. Vessels can be made of ceramic, glass, metal, plastic, or a composite resin material. Ceramic, glass, and plastic usually do not harm or interfere with cells and chemical reactions used in biotechnology. However, they are very fragile materials and must be reserved for small bioreactors.

Larger bioreactors must be made of a stronger material such as metal. Most cells and biological reactions are inhibited by metals. So, metal bioreactors are usually made of stainless steel because they do not corrode or rust if damaged. Corrosion and rusting will leak metals into the contents of the bioreactor. Other metal bioreactors are lined with ceramic or glass to provide stretch and safe conditions in the vessel. Composite resin bioreactors are usually made of fiberglass held together with a plastic resin that does not interfere with the cells or chemical reactions. They can be produced in a variety of shapes and sizes. They are used for a variety of purposes.

It is very critical that the vessel is maintained as a clean and safe environment for carrying out the bioprocessing in the vessel. This is partially accomplished by strict procedures for sterilizing and decontaminating the vessel. Sterilization involves removal or destruction of all microorganisms that can disrupt the bioprocessing. Decontamination is the removal of harmful chemical substances that interfere with bioprocessing. The safe environment inside the vessel is the job of the other bioreactor components.

A continuous motion of the liquid inside the bioreactor is essential to keep the cells or chemicals in the vessel from settling to the bottom.

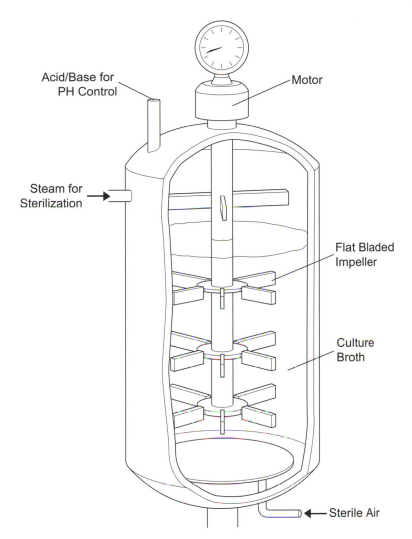

Figure 3.2 Bioreactors are commonly used in biotechnology industries to produce commercial chemicals, food ingredients, and drugs. They are designed to keep cells and microorganisms alive and reproducing. (*Jeff Dixon*)

Settling can inhibit or kill the cells and will slow down chemical reactions that carry out the bioprocessing. Mixing also makes sure that the contents in the vessel are uniform. Uniformity in vessel ensures that cells will get the atmospheric gases and nutrients they need to survive. It also permits chemical reactions to take place at their fastest rate. Mixing can be achieved by rotating or shaking the vessel or by stirring the contents with a propeller. Rotating and shaking is more effective for smaller bioreactors. This type of mixing is difficult in large reactors and does not ensure uniformity in large volumes of liquid. Propellers are used to mix the contents of medium and large vessels.

Mixing must be done very carefully to ensure a uniform distribution of cells or chemicals in the solution without destroying the contents by motion called shear. Shear is a force that distorts and stresses materials being mixed in a solution. Cells and biological molecules are easily destroyed by too much shear. Most modern bioreactors have computer-operated mixing devices that monitor and control shear. Temperature control is equally as important as the mixing process. Too low a temperature will inhibit the function of cells and will slow down the chemical reactions used in bioprocessing. High temperatures can kill cells and destroy the molecules needed for the bioprocessing reactions. Temperature can be controlled with special coils that heat or cool the inner surface of the vessel. Some vessels have coils inside the chamber of the vessel. Mixing is critical to temperature control because it ensures a uniform distribution of temperature within the vessel.

The atmosphere supply of the bioreactor provides the correct atmospheric gasses needed to carry out the bioprocessing. Most cells used in bioprocessing need large amounts of oxygen in order to carry on the metabolism they need for the bioprocessing activities. In contrast, fermentors require low levels of oxygen. Plant cells grown in bioreactors benefit more when maintained in high levels of carbon dioxide and oxygen. Many chemical reactions in bioreactors are inhibited by oxygen. These bioreactors are sometimes provided with an atmosphere high in nitrogen gas. The nitrogen gas is harmless to the bioprocessing and displaces any oxygen that may enter the bioreactor.

Media components such as nutrients and chemicals needed to maintain the conditions for the bioprocessing are added through the media supply system. Media is defined as the chemical components making up the liquid portion of the bioprocessing conditions. The type of media added to a bioreactor is dependent on the types of cells being grown. Bacteria and fungi are usually simple to grow. They mostly require simple mixtures of carbohydrates and proteins that they use as food. Animal and plant cells need chemicals called growth factors as well as precise mixtures of food. Growth factors maintain the normal metabolism of the cells. The pH of the medium is also adjusted using chemicals mixed in through the media supply. Cells and chemical reactions have an optimal pH range needed to carry out the correct type of bioprocessing reactions. In addition, certain chemicals are added to reduce the build-up of waste products made during the bioprocessing reactions.

The collection port as the name implies allows the bioprocessing products to be collected. Collection of the products can be done by draining the whole vessel after a certain period of time. Materials from the bioreactor can also be collected continuously. In a continuous collection system, the other components in the vessel must be returned so that the bioprocessing reactions can continue. Collection ports can also be modified to remove wastes that can inhibit the progress of the bioprocessing. The products made in a bioreactor are a composed of a complex mixture of chemicals. This necessitates the use of other equipment such as centrifuges, chromatography, and filters to purify the products.

The control panel is the heart of the bioprocessing setup and is used to adjust the various components of the bioreactor. Older bioreactors have manual control panels operated by switches and valves that control the atmosphere supply, collection port, control panel, media supply, mixer, and temperature. A system of gauges alerts the operator to the conditions in the vessel. These systems are difficult to monitor and the accuracy of maintaining the process is dependent on the attentiveness and skill of the operator. Newer bioreactors are automated using a computer that monitors and controls the different components and conditions. The operator is mainly responsible for programming the conditions in the vessel. These setups can rapidly respond to changing conditions in the vessel and are capable of making quick adjustments. They can also operate consistently twenty-four hours a day.

Blotting Apparatus

Blotting is a general term used for collecting certain types of DNA, RNA, or proteins in a concentrated sample. A blot is a spot of chemical typically attached to a paper-like material called a membrane used to isolate the sample. Sometimes the blot is referred to as a dot in what is called dot blotting. In general, blotting involves the following steps. In the first step of blotting the sample being studied is separated from other materials in a mixture using a separation technique called electrophoresis. As part of electrophoresis procedure, the sample ends up trapped in a material called the gel. Further analysis of the sample cannot be done because the gel is too thick for carrying out chemical tests on the sample.

One goal of blotting is to extract the sample from the gel and place it on the surface of membrane where chemical analysis can be done. Consequently, the sample is transferred to a membrane that attracts specific chemical components of the mixture. This membrane is composed of a material called nitrocellulose. Nitrocellulose is a special type of paper that attracts and binds to molecules such as carbohydrates, DNA, RNA, and proteins. There are two methods used in transferring the sample from the gel to the membrane. A passive method uses a device that presses the electrophoresis gel onto the membrane. The membrane attracts the chemicals from the electrophoresis gel binding them up tightly to its surface. Another type of blotting uses an electrical current to transfer the chemicals from the electrophoresis gel to the membrane.

The final step involves identification of the desired chemical using a compound called a probe. Probes are specifically designed to bind to the desired chemical somehow making it conspicuous on the membrane. A group of probes called visible probes make the particular chemical glow or appear blue under special conditions. Radioactive probes are used to expose an image of the chemical on photographic film. Scientists can then remove the desired chemical from the membrane once it is identified with the probe. The chemical can be studied further or used in other biotechnology techniques. In 1975, Edwin M. Southern developed the Southern blotting technique to separate and probe desired segments of DNA. The technique, which was named in his honor, used probes made out of DNA. These probes were specifically designed to bind or hybridize to the desired segment of DNA. Southern blotting is used today to identify and locate particular genes in large segments of DNA. Northern blotting uses a similar strategy to find particular segments of RNA. It was named Northern blotting as a pun on Southern's name. DNA or RNA probes can be used in Northern blotting. The identification of proteins can be done using a blotting technique. This type of blotting was called Western blotting. The designation Western blotting kept with the humorous naming convention. Western blotting probes are usually made of antigens designed to bind to a specific protein. As expected, there is a technique called eastern blotting used to identify complex carbohydrates associated with cell structure. Antibodies and other types of probes that adhere to specific carbohydrates are used in this technique.

Centrifuge

Centrifuges of various types are a common sight in biotechnology research laboratories and production facilities. The centrifuge is a machine that rapidly spins liquid samples and separates out various components of the sample by differences in their density. Density is a measure of how heavy a solid, liquid, or gas is for its size or volume. Centrifuges provide a type of work called centrifugal force. Centrifugal force is produced by a rotational movement that moves materials in solution away from the center of rotation. It is the opposite of centripetal force, which is an inward force that keeps the material on a curved path. Centripetal force will interfere with the separation procedure by settling some of the sample to the sides of the container. This effect is usually minimized by placing the sample containers at precise angles that encourage most of the settling to take place vertically in the sample containers.

Centrifugal force has been known to be a good way to separate different chemicals in solution. It is regularly used to separate DNA from other biological molecules that make up cells. Scientists can use centrifugation to collect pure samples of DNA for genetic studies and genetic engineering research. Centrifugal force is also useful for separating impurities from solutions that will be made into biotechnology products. Many bioprocessing operations use centrifuges to remove cells grown in large volumes of liquid as a way of isolating useful chemical secreted by the cells.

Different materials need a different rate of spinning to obtain adequate separation. So, all centrifuges can be adjusted to control the rate at which the sample spins. Spinning can be measured as revolutions per minute (rpm) and as gravitational force units (g-force). The term rpm refers to the number of times that sample completes 360 degree rotation in one minute. The centrifugal force of the spinning produces the measurement called g-force. G-force refers to unit of force equal to the force exerted by gravity. Spinning a sample at a higher rpm produces a greater g-force. Many centrifuges are regularly operated at 10,000 rpm for many biotechnology procedures. Special centrifuges called ultracentrifuges can exceed 100,000 rpm. The g-forces in a sample spinning at 10,000 rpm can exceed 17,000 g-force units. This is the equivalent of a 150-pound person being pressed upon by a 1,275-ton weight. Ultracentrifuges can exceed 1 million g-force units.

All centrifuges have three main components: the sample holder, the spinning device, and the speed control. Sophisticated centrifuges may have additional features such as brakes, refrigeration and heating units, and vacuum pumps that permit them to carry out specialized tasks. The most commonly used centrifuges use a fixed volume sample holder. Fixed volume sample holders are adapted for carrying test tubes or other special containers designed for separating chemicals. Laboratory fixed volume containers can hold microliters of solutions. A one gallon container can hold almost 4 million microliters. Large industrial centrifuges have containers that can hold liters of solution. Continuous flow centrifuges are designed to spin a stream of sample flowing into the spinning device. It separates and collects liquid or solid components of a sample. Industrial continuous flow centrifuges used in bioprocessing operations can process thousands of liters of sample in an hour.

The heart of all centrifuges is the spinning device. Fixed volume centrifuges use either a pivot arm or a rotor attached to a rotating motor. In pivot arm centrifuges, the pivot arm is attached at one end to the motor and at the other end to a sample container holder. The centrifuge motor spins the pivot arm at a high speed placing centrifugal force on the sample in the sample container holder. Most holders are designed such that the container spins in a fully horizontal position. This makes the separation more uniform by producing equal layers of separated components. Denser and heavier components of the sample end up on the bottom of the container. Liquid components usually float to top while solid materials settle to the bottom. A rotor is a disk-shaped holder with openings for placing the sample containers. The sample containers sit at an angle in the rotor so that the spinning of the rotor produces uniform settling in the container.

Continuous flow centrifuges use a hollow rotating drum to hold and separate the sample. Many of them resemble large washing machines. Liquid sample is pumped through a pipe into the drum as the drum is rotating. The rotating action of the drum instantly separates the sample components based on density. Outflow collection pipes are attached to a casing around the drum to gather and transfer the different components to collection chambers. These centrifuges are regularly used in bioprocessing operations.

Chromatography

The term chromatography is literally translated into "making a graph of colors." Traditionally, it was a chemical analysis technique that separated a mixture of chemicals into the separate components that were identified by their different colors. All chromatography techniques have two parts or phases involved in separating the components of chemical mixture. One part is called the stationary phase and the second component is the mobile phase. The stationary phase or immobile phase is designed as a barrier to selectively slow down or accelerate the movement of different chemicals in the mixture. It is very much like running an obstacle course. It can be composed of paper-like material or ground glass sprayed onto a sheet of glass or plastic. Certain molecules traveling across the stationary phase move faster or slower depending on their ability to pass by the obstacles. The mobile phase is either a liquid or a gas that pushes the mixture across the stationary phase. Mobiles phases

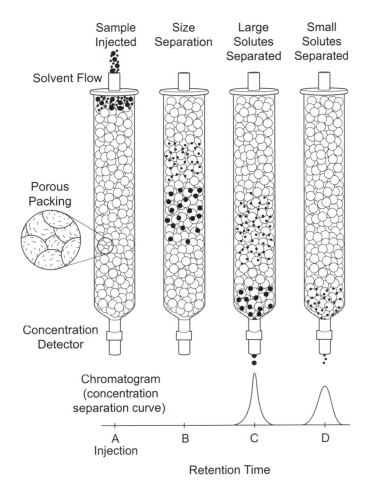

Figure 3.3 Chromatography is a method of separating components of chemicals from a mixture. It is used to study the characteristics of a chemical or can be used as a means of purifying chemicals that have uses in biotechnology applications. (*Jeff Dixon*)

are designed to separate particular types of mixtures. Certain components of the mixture dissolve better in the mobile phase and therefore travel faster as they pass along the stationary phase. Precise combinations of stationary phases and mobile phases are used to separate and identify particular components of a chemical mixture.

The type of "obstacles" designed into the stationary phase depends on the types of chemicals being separated. There is no typical stationary phase. The stationary phase can be a piece or paper or glass coated with a surface covered with the "obstacles." This is called paper or thin layer chromatography. Another type of stationary phase is made of beads with the obstacles bound to the surface of the bead. Beads can be composed of a gel-like material or a glass-like material called silica. Gel materials are soft and used in low pressure chromatography. The hard silica beads are used in chromatography systems in which the mobile phase is passed along the beads at a high pressure. This pressure would crush the soft gel beads.

Each different type of stationary phase helps in defining the specific type of chromatography. For example, stationary phase differences distinguish the following commonly used types of chromatography separation: affinity, chiral, gel permeation, ion exchange, reverse phase, and size exclusion. Affinity chromatography uses chemicals called ligands that temporarily attach to particular molecules. Ligands can be made of antibodies, carbohydrates, enzymes, and other organic molecules. It separates chemicals in a mixture by selectively slowing down the progress of molecules attracted to the ligand. Chiral chromatography uses a stationary phase that separates nearly identical molecules based on very subtle differences in shape. It uses a ligand that attaches to one shape and not the other. Gel permeation uses a special bead. It forces small molecules into the bead causing them to slow down while large molecules glide along unobstructed.

Ion exchange chromatography uses electrically charged beads to slow down the progress of oppositely charged molecules. Hence, a stationary phase with positively charge beads slows down negatively charged molecules letting molecules with a positive charge pass along quickly. Reverse phase chromatography uses specially coated beads or paper that attracts uncharged molecules. This behaves the opposite or in reverse of typical chromatography that uses some type of electrical charge. Thus, in this situation charged particles pass along quickly while uncharged molecules move slowly in the stationary phase. This method is used to separate molecules that are likely to dissolve in fats. Size exclusion chromatography is the simple way to separate a mixture of chemicals. It uses a stationary phase that obstructs large molecules while letting smaller one pass readily along.

As mentioned earlier, the mobile phase provides the push that moves molecules along the stationary phase. Liquid chromatography, as is evident in the name, uses a liquid called a solvent to move the molecules in the mixture. In low pressure liquid chromatography, the solvent drips down the stationary phase moving the mixture slowly across the paper or the beads. A powerful pump is used to move the solvent at speed in high pressure liquid chromatography or HPLC. Gas chromatography uses a high pressure gas mobile phase to move liquids through hollow metal coiled tubes filled with stationary phase. Chromatography can be done using very small amounts of stationary phase in narrow tubes called capillaries. The mobile phase is moved through the capillary by an electrical charge. It is used for rapidly separating and identifying small amounts of molecules in a mixture. This has proven very successful in laboratories developing a variety drugs and medications.

Chromatogram Scanner/Densitometer

Thin layer chromatography is a chemical analysis technique that separates a mixture of chemicals into the separate components identifiable by their pattern of separation. The result of the separation is called a chromatogram. Each band represents a different chemical component separated based in its movement along the material making up the chromatogram. The material, or stationary phase, is usually composed of paper-like material or ground glass sprayed as a thin layer onto a sheet of glass or plastic. Hence, the name thin layer chromatography. A flowing solvent called the mobile phase provides the force that moves molecules over the surface of the stationary phase. Interpreting the chromatogram could be quite tricky if the bands are close. First, inaccuracies in measuring the separated bands are common if done using a pencil and ruler. Moreover, the amount of material present in a band is very difficult to determine. The relative amount of chemical can be calculated by observing the size and intensity of a band. However, the approximate size and intensity of a band cannot be consistently determined just by using a ruler and a person's judgment.

Chromatogram scanners, or densitometers, were designed to read the separation and intensity of bands on a chromatogram. Densitometry is best defined as the measure of the concentration or density of a material such as a spot of chemical. Chromatogram scanners look like larger versions of the document scanners used with computers. The scanner shines a beam of light on the chromatogram and records the image of bands. This image is then fed through a computer program that determines the different degree of separation for each band. The image recorded by the densitometer replaces the traditional drawings and photographs used to record chromatograms.

Not just any type of light is used by the chromatogram scanner. The instrument uses specific ranges of pure light that is either absorbed or reflected by the chemicals in the band. It can be adjusted to use specific types of ultraviolet, visible, or infrared light. Ultraviolet light is most commonly used in fluorescence mode. Fluorescence means to glow. Certain chemicals glow or fluoresce when exposed to ultraviolet light. The chromatogram scanner can measure the fluorescence of a particular band as a way of measuring the amount of chemical in the band. A chemical's concentration or quantity in the band can be calculated by the degree of fluorescence. Visible and infrared light is used for absorption mode. In absorption mode the light taken in or absorbed by the chemical is measured. Specific types of light are absorbed by different chemicals. A built-in computer can calculate the best type of light that gives the most accurate measurement for each band. The scientist operating the instrument can determine the light measurement manually in certain procedures. This is done when the scientist is looking for a particular chemical component that is identified by a specific type of light.

Cryopreservation Equipment

Cryopreservation is described as the process of storing biological samples or whole organisms at extremely low temperatures often for long periods of time. One use of cryopreservation in biotechnology is for shipping genetically modified cells to other laboratories. Scientists who work with agricultural animals commonly use cryopreservation to store fertilized eggs that will later be placed in a female animal. Sperm and unfertilized eggs are commonly placed in cryopreservation equipment in human fertility clinics. The earliest cryopreservation was performed on human sperm in 1776 when it was shown that sperm can survive freezing. In 1938, sperm was shown to survive subfreezing temperatures as low as −269°C and was capable of being stored for long periods of time at −79°C. The first commercial cryopreservation operations were founded in 1972 with the birth of modern biotechnology. Cryopreservation equipment has been greatly improved and refined since then.

Two pieces of equipment are needed to carry out cryopreservation. The first is a vitrification device. Vitrification is a process where cells are rapidly cooled in a manner that prevents ice formation in cells. Cells comprise large amounts of water so that they form ice crystals when they freeze. These ice crystals kill a cell by destroying the delicate structures within the cell. Vitrification is the heart of cryopreservation because it begins the freezing process and must be done properly so as not to damage or kill the cells. The devices that carry out vitrification are composed of a special low temperature freezing unit, a control panel, and specimen holding chamber.

The vitrification freezer is operated to bring the cells from their growing temperature of usually 37°C to a variety of temperature ranges spanning –20°C to –140°C. The freezing process is done in a two-step manner that has been shown through many studies to be safe for cells and whole organisms. A typical freezer can be programmed to freeze a sample from 5°C to –40°C at –1°C per minute and after a short pause from –40°C to –85°C at –4°C per minute. The freezing process is considered rapid because it can be carried out in less than one hour. A manual or computer-operated control panel regulates the temperature change. The control panel is hooked up to an electrical circuit that operates a pump-driven freezer. A special refrigerant liquid is pumped to the specimen holding chamber. There are various types of freezing units that use either fluorocarbons or ethylene glycol refrigerants. Fluorocarbons are used in household freezers and ethylene glycol is found as a coolant in automobile radiators.

The vitrification unit's specimen holding chamber comes in a variety of styles and sizes based on the types of specimens undergoing cryopreservation. Most research laboratories use small units to freeze tubes placed in long narrow tubes or miniature bottles. Larger units are used to freeze big volumes of cells or whole organisms. Biotechnology manufacturing companies are likely to have very large cryopreservation facilities to handle liters of cells produced for commercial sale. Specimens placed in the chamber must be soaked in a special cryopreservation fluid before beginning the freezing process. Cryopreservation fluids are selected based on their freezing properties for a particular type of specimen. These fluids reduce ice crystal formation in the cells and also reduce damage to the cells during the thawing process. The chemicals DMSO (dimethyl sulfoxide), ethylene glycol, glycerol, and propylene glycol are commonly used in biotechnology cryopreservation applications.

The next component of the cryopreservation setup is the storage unit. A typical storage unit is a container filled with liquid nitrogen. The storage unit is an insulated drum that traps the freezing of cold liquid nitrogen. Nitrogen is normally a gas. However, when it is compressed to liquid, the nitrogen drops drastically in temperature. Liquid nitrogen can get as cold as –196°C. Many cells are stored at temperatures from –78°C to 120°C. A cloud of water vapor appears when the storage container is opened because the liquid nitrogen is immediately converted to a cold gas that freezes the water in the atmosphere. Special deep freezers have been developed for holding cryopreservation specimens. The cells must be stored under special conditions to keep the freezers from dehydrating the specimen. Ultra-cold temperatures in the freezer dry out the air holding area making it possible to evaporate the frozen contents from the cells. Dehydration is unlikely to happen in the wet environment of the liquid nitrogen.

Many biotechnology companies are able to make liquid nitrogen on their facilities because the storage units lose large amounts of liquid nitrogen every time they are opened. Even a closed store unit must release liquid nitrogen to prevent an explosion as the liquid nitrogen warms and expands into a gas. Samples removed from the storage units must be thawed in a water bath under certain conditions to prevent damage to the cells. Improper thawing will cause ice crystals to form in the cells. It is not unusual to thaw small specimens for a short period of time in a 37°C water bath and immediately chill on ice until used or processed further. Certain biotechnology use special thawing units that thaw the cells under precise conditions. Thawing is very difficult for larger specimens and requires special treatments that ensure all the cells are not damaged during the thawing process.

Cytometer

Cytometry is a method of counting cells using an instrument called a cytometer. Biotechnology applications that work with various types of cells use cytometers to keep track of the numbers and types of cells used in a process. A specialized type of cytometer called a hemocytometer is used in biotechnology applications involving blood cells. Cytometry was traditionally carried out using a microscope and a special cytometry slide. The slide has a grid engraved onto a surface where a specific volume of liquid is held. When viewed under a microscope, it is possible to count the cells that overlap the grid. A scientist can adjust the magnification of the microscope for identifying the different types of cells. Certain cytometer slides are designed with small scales so that scientists can measure the size of a particular cell.

A flow cytometer is a sophisticated instrument for counting cells. It also allows researchers to determine various characteristics of cells. Some flow cytometers have the capability of separating cells from a mixture of cells based on characteristics determined by the scientist. This ability provides scientists with a simple means of isolating diseased cells or particular genetically modified cells from an assortment of cells. Automated flow cytometers can sort, count, and identify cells at a rate of 500 to 5,000 cells per second. This far exceeds the rate of scientists using a hemocytometer. It is estimated that a skilled scientist can only hand-count cells at a rate of 200 cells per minute.

A typical flow cytometer is composed of a reservoir, laser source, focusing system, detector, and cell sorter. The reservoir forces the cells to flow into a single line of cells down a narrow tube. Certain flow cytometers called capillary cytometers use a very narrow tube to force cells into a narrow band. A control unit permits the user to adjust the rate of flow based on the concentration and types of cells being analyzed. Each cell then passes through a clear portion in the tube called a window. The window is aligned so that the laser light passes through each cell traveling past the window. Different types of cells require a particular color or wavelength of laser light for identification and counting. This means that flow cytometers are usually designed to count or identify a particular type of cell. Certain types of cytometers have multiple lasers that give the researcher the versatility of analyzing different types of cells.

The cytometer is able to count cells because the detector is able to determine the presence of a cell when the laser beam going to the detector is interrupted. Cell size is determined by a feature called forward scatter. Forward scatter refers to laser light that bounces or is diffracted around the cell. The amount of forward scatter is proportional to the size or volume of the cell. A feature called side scatter is related to the internal complexity of a cell and is useful for identifying different types of cells. Cell identity is also assisted by using different color lasers to determine unique characteristics of a cell. The focusing system is designed to help the detector collect the scattered light. Scatter patterns are then determined by a computer linked to the detector.

Certain flow cytometers have a separating device called a cell sorter that places cells into separate containers based on size or other characteristics. Sorting is usually achieved using a sorting nozzle. A computer controls the position of the nozzle over a series of collecting containers sitting at the end of the reservoir tube. The computer is able to take the information collected by the detector for identifying particular characteristics of a cell passing through the window. This information then controls a small robot that moves the nozzle over a container corresponding to the characteristics. Scientists can then use the cells for further analysis or for research studies.

DNA Sequencer

DNA sequencers permit scientists to determine the nucleic acid sequence of a length of DNA. This provides valuable information for genomic researchers investigating the identity of genes. It permits them to compare similar genes of different animals. The technology also provides information about the differences between normal and defective genes. There are two types of DNA sequencer technology. The older or traditional method uses a special polyacrylamide electrophoresis procedure. First, one strand of the DNA is exposed. Chemicals called primers are then added to the open DNA strand. Varying sized copies of the DNA are then made. These fragments are then labeled with radioactive elements. Each fragment is labeled in such a way that the researcher knows the nucleic acid located at one end of the fragment. This is done with a radioactive marker that selectively sticks to the particular base of the fragments. The labeled fragments are then placed on a large thin gel made up of a material called polyacrylamide. Four columns of fragments are run. Each column represents a fragment with one of the four nucleic acids (A = Adenine, C = Cytosine, G = Guanine, and T = Thymine). An electric charge is then passed through the gel for a set amount of time attracting the negatively charged DNA fragments to the positive electrode or cathode at the bottom of the gel. Smaller DNA fragments travel more quickly through the gel ending up on the bottom. The gel is then placed on a large X-ray film that shows the fragments as dark spots wherever the radiation exposes the film. It is then analyzed by a scientist or by an instrument called a gel reader. Gel reading can be very difficult and is subject to many errors. So, the gels must be read at least twice to ensure accurate interpretation of the nucleic acid sequence.

A new method has been devised to sequence DNA. It is simple to use and does not require the dangerous and difficult-to-dispose radioactive labels. In addition, it is integrated into a computer system that eliminates the need for the standard gel interpretation method. It starts out just like traditional sequencing because the DNA is replicated into differing size fragments ending in each of one of the four nucleic acids. However, it varies after this point. One end of each fragment is labeled with a special dye that specifically attaches to one particular type of nucleic acid. The dye is not radioactive. Rather it is a special dye that glows a specified color when exposed to the light of a laser. These are called laser activated dyes. The fragments are collected by a tube that feeds the fragments through a column that separates each fragment based on size. A laser shines through a clear opening in the film causing the dyes to glow the specified color for each nucleic acid as the fragments pass along. This information is recorded as a chart that calculates the nucleic acid sequence. The readout is much more accurate than the traditional sequencing method.

Electrophoresis

Many scientists consider electrophoresis as the workhorse of biotechnology. It was one of the first simple technologies developed to analyze nucleic acids and proteins. As its name implies electrophoresis uses electricity (electro) to transport (phoresis) particles. Scientists discovered that different particles move through an electric field based on their charge. The idea of using electricity to separate biological molecules came from Swedish biochemist Arne Tiselius. He was awarded the 1948 Nobel Prize in chemistry for this and other biochemical separating technologies. His discovery made it much simpler to study nucleic acids helping advance the newly created field of molecular genetics.

The U.S. Department of Energy Human Genome Project Information Center defines electrophoresis as "A method of separating large molecules from a mixture of similar molecules. An electric current is passed through a medium containing the mixture, and each kind of molecule travels through the medium at a different rate, depending on its electrical charge and size. Agarose and acrylamide gels are the media commonly used for electrophoresis of proteins and nucleic acids." Electrophoresis is most commonly used to identify DNA fragments and whole proteins. It is usually followed up with a technique called blotting to specifically identify a particular DNA segment or certain type of protein.

Traditional electrophoresis uses an electric current to push electrically charged biological molecules through a porous solid material called a gel. Agarose is a jellylike polysaccharide used in one type of electrophoresis. It is commonly used as a thickening agent in cosmetics, drugs, and food. The agarose is heated in water and allowed to cool in a chamber that molds the gel into a flat horizontal slab. It has holes called wells cut into the gel. These wells hold the samples that are going to be separated. DNA, RNA, and proteins are separated using agarose gels.

Polyacrylamide is the other common electrophoresis gel. It is made by mixing an organic chemical called acrylamide with a catalyst. This causes the acrylamide to bind to itself forming polyacrylamide. Polyacrylimide is commonly used as a thickening agent in cosmetics and plastics. Moreover, it is used in water treatment and as a soil-binding agent to prevent erosion. The mixture then hardens into a porous gel. Acrylamide is a neurotoxin and may cause cancer in people. Hence it is handled very carefully. It is somewhat safe in the polyacrylamide form. Polyacrylamide is molded into vertical slabs with sample wells notched out of the top. Proteins are usually separated on polyacrylamide gels.

Gell Electrophoresis

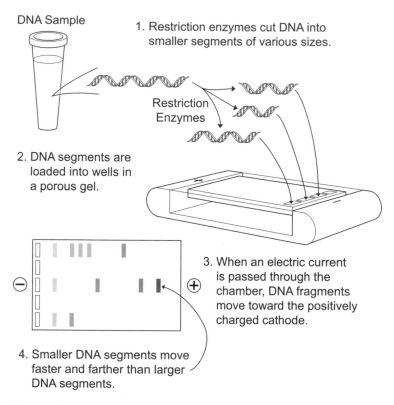

DNA Sample

1. Restriction enzymes cut DNA into smaller segments of various sizes.

Restriction Enzymes

2. DNA segments are loaded into wells in a porous gel.

3. When an electric current is passed through the chamber, DNA fragments move toward the positively charged cathode.

4. Smaller DNA segments move faster and farther than larger DNA segments.

Figure 3.4 Electrophoresis is commonly used in many biotechnology laboratories to separate samples of DNA, RNA, and proteins for analysis and purification. (*Jeff Dixon*)

The electrical current in electrophoresis is the primary driving force separating molecules in a mixture. Biological molecules are usually placed in a solution that accentuates their negative charges. The negatively charged molecules in the mixture are attracted to the positive electrode or cathode of the electrophoresis chamber. This driving force pushes the molecules through the gel. Samples in agarose gels are separated based on their relative sizes. Large molecules do not move as quickly through the gels as smaller molecules. Thus, small DNA particles are found closer to the cathode. The movement of proteins in polyacrylamide gels is more complex. Standard polyacrylamide gels (PAGE) separate the mixture down the gel based on differences in the protein's degree of electrical charge, shape, and size. A type of treatment called denaturation is used in SDS gels. The proteins are heated in a soap solution called sodium dodecyl sulfate or SDS. This causes the proteins to have similar charges and shapes. Therefore, separation is based on size. Again, smaller molecules move more quickly to the cathode on the lower portion of the gel.

There are many variations to electrophoresis. One variation is called two-dimensional electrophoresis. This is usually used for separating proteins. It permits better isolation of proteins with similar characteristics. In this method, the proteins are separated using a particular PAGE or other procedure. This separated sample is then placed in another setup at a 90 degree angle to the original. In the second setup the sample is run through an SDS system. Another variation is called capillary electrophoresis. In this method of separation, the gel is replaced with a very fine tube coated with a surface that permits the smooth passage of the proteins flowing in a liquid. Smaller samples of proteins can be separated more distinctly using this method. Since there is no gel, a spectrophotometer detector must be used to record the samples as they separate from the mixture.

Electroporation Instrument

Electroporation is one of several techniques used for introducing DNA into a cell for genetic engineering. The instrument is simply a system for delivering a precise amount of electrical current into a liquid culture of cells. It is not common to use currents exceeding 250 volts to carry out the electroporation technique. This is over twice the voltage that runs through the average electrical outlet in a house. Electroporation is based on the principle that cells grown under certain conditions can take up pieces of DNA when exposed to a particular electrical charge. Cells are grown in a special fluid or medium that prepares the cell for genetic engineering. They are then mixed with a specifically processed piece of DNA containing a desired gene. The cells and DNA are then subjected to a particular electrical treatment. Cells produce "pores" in their membranes when exposed to electrical currents under certain conditions. This is where the term electroporation was derived. These pores then permit passage of the DNA into the cells. The cells are then tested to see if they are using the newly inserted genes.

The electroporation instrument can be used for other techniques that require modification of cell membrane properties. Researchers can adjust the settings on electroporation equipment in a manner that permits the fusion of one cell to another. This technique of using electricity to fuse cells is called electrofusion. Electrofusion can be used to fuse two

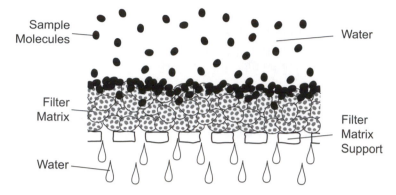

Figure 3.5 Filtration is an important means of collecting or separating biotechnology products made in bioreactors. The filter permits the scientist to collect the product from a solution. (*Jeff Dixon*)

similar cells having dissimilar genetic material. This strategy instantly produces a cell with combined characteristics. Recent studies are blending cancer cells with white blood cells that produce protein antibodies. These cells, called hybridomas, are used to produce biotechnology vaccines that can even be applied to protect people from cancer. Scientists are currently trying to use electroporation as a method of combining the sperm and eggs of unrelated organisms.

Filtration Apparatus

Filtration is a common approach used for separating solids from gases or liquids in many biotechnology applications. The process of filtration passes gases or water through one or more layers of a porous material called the filter matrix that traps solid particles. A filter matrix can be composed of various categories of materials. The most common filters are made of ceramic, charcoal, cloth, glass, organic polymers, plastic, paper, and sand. Filters are mostly used to remove solids of a particular size from a solution. However, filters can also be used to separate solids based on their chemical properties by their electrical charges.

Each category of filter matrix has a set of characteristics that make it favorable for separating a particular type of chemical from a mixture of chemicals in a gas or liquid. For example, glass filters are favorable for separating chemicals that can corrode paper or plastic. Organic polymer filters can be designed to separate mixtures based on the chemical or electrical properties of the chemicals in the gases or the solution. Cloth filters are useful in working with fragile solids used in biotechnology applications. The solids that are captured by the filter can be collected without damage by gently soaking the filters or shaking the solids loose using vibration or a stream of air.

Another important characteristic of filters is a property called porosity. Porosity is defined as the percentage of open spaces, or pores, found throughout a filter matrix. These open spaces in the filter permit the liquid to pass through the matrix while trapping the solid materials. Porosity helps in determining the shape and size of the particle that is trapped in the matrix. It also affects the rate at which the liquid flows through the filter. Porosity is characterized by the shape, size, and distribution of the open spaces. The filters commonly used in biotechnology applications have the following porosity types: capillary membrane, fabric, fibrous, and porous membrane.

Capillary porous membrane filters have matrix with cylindrical pores of uniform diameter that run perpendicular to the filter surface. They are composed of organic polymers, plastic, or special paper-like materials. These filters are useful for capturing particles of a specific size. They are designed so that the gas or liquid easily passes through the pores. Particles that are smaller than the pores also pass through the filter matrix while the larger particles remain trapped on the upper surface of the filter. These filters are commonly used to separate hazardous chemicals from water used in many biotechnology applications. Capillary filters have medical applications too. They are currently used in artificial kidneys that separate certain wastes from the blood while retaining beneficial blood components such as cells and proteins.

Fabric filters, as the name implies, are made of cotton fabric similar to what is found in clothing and other textiles. These filters are most useful in removing contaminants from the air in biotechnology facilities. They are particularly important in areas called clean rooms that must have pure air that is free of any chemicals or microorganisms that would contaminate cells or bioprocessing operations. Large fabric filters are used in bioprocessing called filter presses. Filter presses are commonly used to collect biotechnology products from the liquids in bioreactors. Special fabric filters that remove very small materials from air are called HEPA or high efficiency particulate air filters. Many types of vacuum cleaners and air purifying fans are being designed with HEPA filters that remove dust from offices and homes. HEPA filters can also be found in dust masks and face masks commonly used in biotechnology operations to protect the workers from hazardous chemicals in the air.

Fibrous filters are often confused with fabric filters because they have the same types of pores. The pores are irregularly shaped channels formed by the fibers woven into the filter matrix. Unlike fabric filters they are composed of a matrix of fine fibers arranged so that the openings in the filter are perpendicular to the surface. The most common types of fibers used in fibrous filters are glass, paper, and plastic. These filters have a high porosity meaning that they rapidly pass gases or liquids through the matrix. Fibrous filters are commonly used to filter large dust particles from air in bioprocessing facilities. These filters also have many applications for removing solid material from water in bioprocessing. Certain procedures use fibrous filters to remove cells from liquids containing biotechnology products made in bioreactors.

Porous membrane filters are also known as screen or sieve filters. They are generally small filters that have specific pore sizes used to remove particles or microorganisms from various liquids. These filters have a large number of pores and are made from strong materials that permit the filtering of rapidly moving fluids or liquids passed into the filter under high pressures. The filter matrix is usually composed of modified paper, organic polymers, plastics, powdered metals, and Teflon®. Porous membrane filters play a vital role in removing contaminants from drugs manufactured by bioprocessing techniques. The filters are also used in product testing. They are attached to the openings of chemical analysis instruments used for various types of chromatography. Large particles that can clog the instruments are removed by these filters.

The size and thickness of the filter matrix varies greatly with the filtering applications. Some filters are very small and thin such as those found in syringe filters. Syringe filters are used to remove contaminants from very small samples of biotechnology liquids being tested using various chemical analysis techniques. Bulky filters are commonly found in industrial Buchner funnels use to collect solids from large volumes of liquid. These filters are commonly used to purify products made in bioprocessing operations. A filtering characteristic called flow rate is affected by the porosity, size, and thickness of the filter. High flow rates are essential for passing large volumes of gases or liquids through the filter. The flow rate of a matrix can be increased by making the filter more porous, larger, or thinner. Small amounts of liquid must be passed through minute filters. This has to be done because a large filter could soak up all of the liquid in the pores. Consequently, nothing will flow through the filter.

The movement of the gas or liquid through the filter matrix can be passive or pressure driven. Passive movement or passive flow is generally used to push liquids through a filter. It uses gravity to push a liquid through the pores very much like the way a coffee filter works. This type of filtration is slow and is usually restricted to small volumes of liquid passed through thin filters. During passive filtration, large or thick filters tend to retain much of the liquid in the pores which is the reason for using smaller and thinner filters. Passive filtration is usually performed by folding the filter and placing it in a funnel. The fluid is then poured onto the upper surface of the filter where it passes through the pores and is then collected by the funnel. A container placed underneath the funnel captures the liquid that has passed through the filter.

Pressure-driven filtration is achieved by forcing a gas or liquid using some type of fan or pump. This type of filtration is used to move large volumes of substance through big filtering systems. There are two types of pressure-driven filtration systems: positive pressure and negative pressure. A positive pressure system forces the gas or liquid through the filter using a fan or pump. Gases are moved with a fan whereas liquids are pushed with a pump. The pressure placed on the gas or liquid is called the head pressure. A particular head pressure must be calculated for the particular substance being filtered and for the type of filter matrix being used. Too low a head pressure can prevent the substances from passing through the filter. A very high head pressure can damage the filter. Negative pressure filtration uses a vacuum pump to suck the liquid through the filter. A head pressure also develops during vacuum filtration. Special funnels that support the filter against the pressure or the vacuum is used in pressure-driven filtration systems.

Gel Reader

When electrophoresis was first developed it was expected that the gel's information was recorded by being drawn on a piece of paper. The information was simply a sequence of bands representing individual proteins or fragments of DNA. Electrophoresis required special viewing because the bands were either faint on clear background or only visible with ultraviolet light. Hence, the bands were drawn while being viewed on a light box that permitted the bands to be seen clearly. However, drawings alone did not provide the detail and consistency needed for the strict precision required of scientific data recording. A mathematical value called R_f, or retention factor, was developed to provide a consistent way to record the location of the proteins or DNA fragments on the gel. R_f is defined as the distance that the center of the band moved divided by the distance the a marker moved. The marker is a dye that indicates how long the electrophoresis separation was running. Both are measured from an established origin. Retention is a measure of the rate at which a substance moves in a chemical separation system. The retention of a molecule varies with the nature of the chemical separation technique being used and the type of molecule being separated. Retention is used as one measure for calculating the properties of a molecule.

Some research required a permanent record of the gel. Scientists working in pharmaceutical companies in particular needed the gel as a document for government recordkeeping requirements. Special ways were developed for drying in preserving gels. However, these techniques distorted the gels and affected the accuracy of the R_f readings. Clever scientists created a way of recording gels without having to draw them. Some placed the gel on a photocopier to try to get an accurate recording. Unfortunately, the copiers in those days did not produce a quality picture worthy of a legal document. Next was the strategy of using document scanners hooked up to computers. This produced high quality images that could be saved to a disk or sent elsewhere by e-mail. Companies that manufactured document scanners learned of their use in electrophoresis and developed gel scanners that linked to better software for producing gel images and gathering more details from the gels. The new software also permits scientists to incorporate the gel images into technical reports and publications.

Gene Gun

The gene gun as its name and appearance imply shoots genes into a cell for carrying out genetic engineering. There are many types of gene guns. One style of gene gun shoots out specific pieces of DNA that are attached to particles of gold. A burst of helium gas sends out the particles at high speed into a culture of cells. Other gene guns shoot the blast of helium through a membrane covering the cells. The membrane contains microscopic polyacrylimide spheres coated with DNA. Polyacrylimide is a gel-like substance in which DNA can be imbedded. These spheres are produced using a new biotechnology method called nanotechnology.

Gene gun techniques are collectively called particle bombardment. It works on the principles that the particles carrying the DNA bombard and puncture the cell's covering depositing the DNA inside the cell. Not all of the cells take up and use the DNA. In addition, some cells take up more DNA than others affecting how the new trait comes out. Consequently, the scientists have to select those cells that show the correct characteristics. They do this with specific cell growth conditions and by measuring any indication of the gene's characteristics. Older gene guns were large instruments that required special handling in order to get consistent results. They produced inconsistent results. Some cells had an overabundance of genes while others had very few. Nanotechnology researchers are currently developing microscopic gene guns that can target one cell at a time to ensure accurate and consistent delivery of the genes. These gene guns can be integrated into robot systems that can handle many cells rapidly and with precision.

Incubator

The concept of incubation refers to the maintenance of controlled environmental conditions needed to sustain the development or growth of cells, eggs, tissues, or whole organisms. A cell incubator is an apparatus used to grow and maintain cell cultures of animals, microbes, or plants. The basic cell incubator keeps cultures under sterile conditions at an optimal temperature and humidity. Many cell incubators also regulate the carbon dioxide, nitrogen, and oxygen composition of the atmosphere in the incubator. Incubators that have special controls for monitoring carbon dioxide are called CO_2 (carbon dioxide) incubators. Many microbes and animal cells require specific carbon dioxide levels to grow normally. Cell incubators are necessary for growing cells used in biotechnology research and bioprocessing.

The principles of cell incubation date back to ancient China and Egypt. People produced simple incubators that used fire to keep the temperatures needed to rapidly develop chicken eggs. Scientists later applied the principle to maintaining microbial cultures used in research. Physicians realized that incubators were useful for studying pathogenic microbes grown under conditions that mimicked the human body. The temperature of these early incubators was controlled using alcohol lamps and electricity. Atmospheric gases were maintained by running chemical reactions that produced carbon dioxide or oxygen gas in the incubator.

Incubators used in modern biotechnology applications are complex devices composed of a growth chamber and a computer-driven environmental control system. The temperature is controlled by precise heating and cooling systems. Atmospheric gases are released at exactly controlled levels from gas storage containers. Light and humidity are also maintained by various devices built into the incubator. Sensors that monitor the environment within the incubator provide constant feedback of the growing conditions. Computers take this feedback to keep the cells at a favorable environment. Typical growing condition for human cells used in biotechnology is 37°C, at 95 percent relative humidity, and 5 percent CO_2 levels.

Cell incubators usually come in one of three types of designs: air draft, dry wall, and water jacket. Air draft incubators circulate air throughout the interior of the incubator to maintain constant temperatures. This style of incubator is used for growing large amounts of cells in a big growth chamber. These incubators are designed to respond quickly to environmental needs of the cells. However, these incubators lose the temperature very quickly and must consistently adjust the air to control the temperature. Dry wall incubators pass air within the walls of the incubator. The walls then radiate the temperature to the growing chamber. For this reason these incubators are sometimes called radiant incubators. They are better at maintaining a constant temperature because the jacket does not let heat escape through the walls of the incubator. Water-jacketed incubators advantages include stable temperature control and increased security in the event of power failure (due to water's natural insulation abilities). Water-jacket incubators are surrounded by water within the walls making up the three sides, the top, and the bottom. These are usually smaller incubators and work by the same principles as dry wall incubators.

Most incubators use infrared light sensors to detect atmospheric gas levels and digital thermometers to monitor temperature. Cell incubators vary greatly in chamber dimensions. Incubators with small incubators are popular in many research laboratories and medical testing facilities. Some companies have made small portable incubators that can be transported from one laboratory to another without disrupting the growth of the cells. Very large incubators are common in biotechnology industries and pharmaceutical companies. Some are large enough to walk into and have dozens of shelves for growing a multitude of cells. Super small incubators are being developed as a means of researching small amounts of cells grown together under a variety conditions. They are proving to be very useful in the development of biotechnology laboratories that develop medical treatments.

Many incubators have built-in alarms that alert of any conditions that can harm the cells. Some incubators can even send a message through the telephone or through the e-mail to warn of potential problems. The control panel of incubators will vary greatly with the types of built-in features. On many of the older incubators, scientists can control the environmental conditions manually by adjusting a knob on the control panel. Newer incubators have computers with programmable software that controls the conditions. Cells grown in the incubators are kept in cell culture containers that are stacked onto shelves within the incubator. Cell culture containers vary greatly in shape and size. They are also made up of a variety of different types of glass or plastic. Scientists must carefully select containers that are appropriate for the type of cell and the growing conditions maintained in the incubator.

Isoelectric Focusing Apparatus

Isoelectric focusing or IEF is a modified form of electrophoresis use for distinguishing different types of proteins. Electrophoresis uses electricity to transport and separate different molecules. Molecules such as proteins move through an electric field based on their charge and other properties particular to a type protein. The traditional gel used in electrophoresis serves as porous solid material that lets the current and proteins pass through it. Its chemistry is uniform throughout so that the proteins travel consistently from the wells to the point where it stops when the unit is turned off. Isoelectric focusing modifies the gel so that it separates proteins based on a property call isoelectric point.

A protein's isoelectric point is defined as the pH where a protein has a charge of zero. The term pH refers to a chemical measure called the potential (p) of hydrogen (H). It is always written with a lower case "p" and upper case "H." Hydrogen potential is a measure of the activity of hydrogen ions (H+) in a solution, which is interpreted as its acidity or alkalinity. Acidity and alkalinity affect the electrical charges on all molecules. Almost all proteins carry an electric charge. This is why they can be separated using electrophoresis. However, their charge can be eliminated at a particular pH specific to the protein. A molecule with its charge does not migrate in an electric field. Hence, a protein at its isoelectric pH will not move when subjected to the normal conditions for electrophoresis.

Isoelectric focusing gels are not uniform in chemistry throughout the gel. The gel is set up so that different segments along the gel have a specific pH. A pH gradient is set up by preparing the gel with chemicals called carrier ampholytes. Ampholytes are chemicals that act as acids or bases and can be used to produce a particular pH. The gel is made in such a way that one end of the gel has a different pH than the other end. This creates a gel with a pH gradient that ranges from acid to base. Proteins migrate through the gel until they lose their charge at their isoelectric point pH. The electric current is not able to move proteins that have lost their charge. This gives a precise way to identify proteins that are difficult to distinguish from other proteins using traditional electrophoresis. This is because each protein has a highly characteristic isoelectric point. It is similar to identifying people by their appearances compared to trying to identify them by their shadows. Isoelectric focusing is combined with other techniques when scientists are working out the characteristics of newly discovered or artificially made proteins.

LIMS

LIMS is the abbreviation given to laboratory information management systems. In simple terms it refers to the computers and software used to handle laboratory data. It is a crucial and ever growing tool in biotechnology. This is because of the great volumes of data collected when analyzing or creating chemicals used in biotechnology applications. A typical LIMS setup uses a central computer or a bank of computers to collect and organize data from various laboratory instruments. The simplest LIMS collects data from laboratory instruments and integrates it into files that can later be used for reports. A scientist can use the LIMS software to look at the data in many formats and select the appropriate type of statistics to analyze their findings. The software can combine the information from chromatography, electrophoresis, and spectroscopy to come up with a complete picture of a chemical's characteristics.

Complex LIMS setups do a variety of jobs depending on the type of laboratory or industrial setting they are operating within. Some industrial and research laboratories use LIMS to control robotic setups that control the laboratory instruments. They add the samples to the instruments, analyze the data, and incorporate the information into a database for investigation later. Industrial systems can control an array of robots and machinery involved in making biotechnology products. Instruments built into the machinery monitor the process and make sure that all the necessary government guidelines or standards are met. Some of the LIMS programs keep track of every chemical or item used in the process ensuring that any problems can be traced to a particular chemical. This also helps scientists keep an eye on inventories so that they do not run out of chemicals and materials needed to carry out the work. LIMS also help scientists look up research studies and government regulations online to help them produce publications and reports. Some systems have helped save companies and research scientists much money by using computers and robots in place of human labor to carry out many routine tasks.

Lyophilizer

Lyophilizers are little instruments found in many type of bioprocessing facilities. The process of lyophilization is the rapid freezing of a solution of chemicals at low temperature followed by dehydration using a high vacuum. This method is used to remove the liquid from sensitive

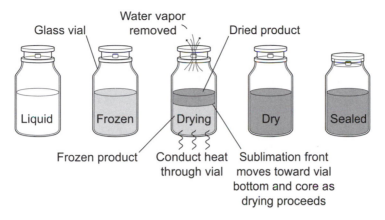

Figure 3.6 Lyophilization uses a four-step process to freeze-dry chemicals. The freeze-dried sample is usually sealed afterward to prevent contamination. (*Jeff Dixon*)

biochemicals leaving behind a fluffy dry powder. The technique dries the chemicals without diminishing their function. Lyophilization is frequently used as a strategy for preserving many bioprocessing products used as pharmaceuticals in medical diagnostic applications. A similar type of lyophilization called freeze-drying is used to preserve the meals ready to eat (MREs) used by campers and the military. Lyophilization ensures that vitamins in the foods are preserved as well as reduces the spoilage of foods that must normally be refrigerated. It is possible to store lyophilized foods for years without refrigeration. However, lyophilized biotechnology products are generally stored cold or frozen to reduce any small amounts of decay that could affect the usefulness of the chemicals.

A typical lyophilizer uses a four-step process to freeze-dry the sample. The chemical sample is dissolved in water and is usually mixed with some type of stabilizing agent such as the sugar lactose that prevents the sample from clumping when lyophilized. Clumped samples are not uniform in consistency and can end up having the characteristics of a lump of melted plastic. This renders the product useless and can destroy the chemistry of the biochemical. The sample separates from any ice that forms during the freezing process. After freezing, the sample is rapidly subjected to a strong vacuum that evaporates the ice leaving behind the dried chemical mixed uniformly with the stabilizing agent. Sublimation is the name given to the evaporation of ice into a vapor that can then be carried away with a pump. The sample can then be sealed and stored for further processing or packaged for its intended use.

Lyophilizers come in a variety of sizes. Small lyophilizers that can fit on a laboratory desk are used in research laboratories and small manufacturing operations. The sample is placed in small jars that are frozen separately in a research freezer or in a freezer build into the lyophilizer unit. A pump then sucks the air out of the jar leaving behind the lyophilized sample. Many scientists use these "bench-top" lyophilizers to store small amounts of biochemicals collected from their research projects. Large lyophilizers are very common in biotechnology manufacturing facilities that produce thousands of liters of material in one operation. The samples in these lyophilizers are poured into containers and placed on a flat tray similar to a cookie sheet. Samples can be poured directly onto the flat trays in certain types of lyophilizers. A built-in freezer than chills the tray and subsequently freezes the solution containing the chemicals. A big pump then produces the vacuum that evaporates and collects the ice. The sample is then sealed in the containers or scraped from the trays and placed into sealed containers.

The high cost and difficulties of operating large lyophilizers provided opportunities for companies that specialize in lyophilization. These companies must maintain strict standards that ensure a consistent product time after time. Special techniques are needed for lyophilizing chemicals used in medical applications. The lyophilization process must be performed in a way that does not contaminate the sample with metals, microorganisms, and organic chemicals. They must also ensure that the lyophilization process does not destroy any of the intended activity of the biochemicals. Many governments have rigid guidelines that regulate the lyophilzation of drugs and foods. Computer systems built into precision lyophilizers provide a record of all the conditions that take place step-by-step during the lyophilization process. This record contains valuable information that can be studied if the product is discovered to be defective after lyophilization.

Microarray Technology

Microarray technology is rapidly replacing traditional ways of identifying the functions of particular sequences of DNA. It is most useful for investigating the interactions between large numbers of genes. Most of the microarrays currently being developed are for characterizing the genetic mechanisms of animal, human, and plant diseases. This technology uses small pieces of artificial DNA to identify DNA sequences being used at a particular time by a cell. Strands of DNA having a specific nucleic acid sequence are chemically attached to a glass slide or a microchip. A microchip is small silicon wafer having thousands electronic components. Microarray slides are designed to give results that can be analyzed visually. Microchips are placed in computers that read and interpret the results. The computer generated results can then be fed into a database to compare the results with other microrray databases.

Microarray technology relies on the ability for scientists to collect the mRNA from a cell expressing a trait of interest. Cells produce a chemical called mRNA when a sequence of DNA is being used to express a specific trait. The cell uses the mRNA to make proteins responsible for the trait's characteristics. Currently, there are many traits for which scientists do not know all of the gene sequences that cooperate to produce the trait. Scientists can use microarrays to identify the genes being expressed by first collecting the mRNA from the cell. The mRNA is then converted into a mirror image or complementary DNA or cDNA strand. They use a technique called reverse transcription polymerase chain reaction (RT-PCR) to carry out this

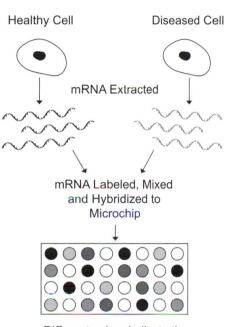

Different colors indicate the absence or presence of a disease gene

Figure 3.7 Microarrays permit scientists to study the differences in gene expression between healthy and diseased cells. (*Jeff Dixon*)

task. The cDNA is then attached to fluorescent labels. This labeled DNA is now called a probe. The labeled probes are added to the DNA strands on the slide or microchip. The slides or microchips are put into instruments that measure the attachment to a known DNA sequence. Each spot on an array is associated with the DNA sequence for a particular gene placed on the microarray. Each color in the array corresponds to normal or diseased expression. The location and intensity of a color indicates whether the cell is expressing normal or abnormal DNA sequences. For example, the green color on a microarray may represent genes expressed for a normal genetic condition. Red represents the diseased condition. The black areas represent no DNA binding. Other colors can be used to identify various expression characteristics.

Microplate Reader

Microplate readers are special instruments designed to measure or monitor up to 96 chemical samples in a single procedure. The samples are placed in special containers called microwell plates. Microwell plates hold small quantities of chemical and can be used to hold up to 96 samples. The wells are ideal for carrying out a variety of chemical reactions at once with a large number of samples. Much of biotechnology involves measuring quantities of molecules in a solution. It is also important to be able to monitor chemical reactions that use up or produce a particular type of molecule. Microplate readers are valuable tools in medical laboratories where they are used to analyze multiple samples of body fluids for disease. In biotechnology labs microplate readers are very important for detecting the action of particular genes being used in genetic engineering studies.

The microplate reader works by shining a particular type of light at each of the samples in microwell plate. It can be adjusted to in various ways. It can read a few to all the samples in a particular sequence or it can read several samples at a time. A particular type of light is selected based on the type of analysis being done. Some chemicals absorb a particular color of light. Their presence or quantity can be determined by measuring how much of the light is absorbed by the sample. This is called absorbance detection. Hence, a scientist looking for the production of a particular chemical made by a cell would notice more and more light being absorbed by the reader as the chemical is produced. Some chemicals glow when exposed to a particular light. This is called fluorescence detection. The amount of chemical is measured by the intensity of glowing. Microplate readers feed the absorbance or fluorescence measures into a computer program that analyses the particular information being collected.

Microscope

Microscopes have made the greatest contributions to biotechnology than any other modern instrument. It was the invention of the microscope that fueled the curiosity to understand the biology behind contemporary biotechnology. The microscopes used in modern biotechnology today vary greatly from the first microscopes used to view biological specimens in late 1600s and early 1700s. These early instruments used simple lens to magnify small objects. Hence they were called microscopes from the terms "micro," meaning small, and "scope," to view. The early microscopic images were not very detailed and did not greatly magnify the object. Some were just slightly better than some of the highest power magnifying lens found selling today.

All microscopes have an illumination source, specimen holder, image focusing devices, and image viewing region. The illumination source is the means of visualizing the specimen being observed through the microscope. Visualize means to form an image. Microscopes are primarily categorized by the type of illumination source. Light microscopes use the light spectrum to help visualize the specimen. Most light microscopes use white light as an illumination source. White light is a combination of all the different colors or wavelengths of light. Precision research microscopes usually use blue light because its small wavelength provides better magnifying power and gives a clearer image. Specialized procedures used in biotechnology require infrared and ultraviolet light sources for illumination. This light is invisible to humans and the image must be viewed using a special screen.

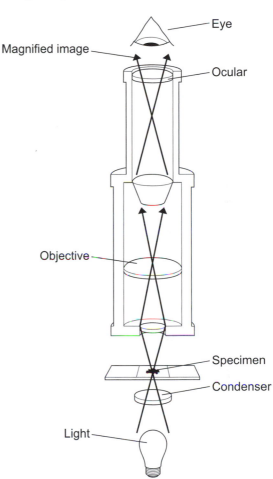

Figure 3.8 The microscope is one of the oldest tools of biotechnology. Specialized microscopes are still used today to study the chemistry, function, and structure of cells. (*Jeff Dixon*)

Electron microscopes use an electron gun as the illumination source. An electron gun produces a beam of electrons to visualize the specimen. Images produced by electron microscopes are not visible to the eye so they must be viewed with a special screen. Electron microscopes are invaluable tools in biotechnology. X-ray microscopes are related to electron microscopes. The illumination source is an X-ray tube similar to those found on the machines used in medical imaging. X-ray microscopes have limited uses in biotechnology and are mostly used to study biomaterials used for manufacturing medical devices. The latest microscope being developed is the nuclear magnetic resonance (NMR) microscope. This microscope uses a radiofrequency generator as the illumination source. A strong magnet interacts with atoms in a specimen to produce a radiofrequency signal that is viewed as an image on a screen. NMR microscopes are finding many uses in biotechnology.

Light, electron, and X-ray microscopes can be used in one of two different types of visualization modes: transmission mode and surface-viewing mode. In transmission mode the image is visualized with the illumination source passing the light, electrons, or X-rays through the specimen. This permits the scientist to see the inside of the specimen. The transmission light microscope is the standard microscope used in school and research laboratories. These microscopes are capable of magnifying images clearly up to 1,000 magnification. The transmission electron microscope is capable of over one million magnification. Surface-viewing light microscopes are also called dissection scopes. The light bounces, or reflects, off the surface of the specimen and is only capable of 200 magnification. Scanning electron microscope is also for viewing the surfaces of specimens. The specimen must be covered with a fine layer of gold in order to reflect the electrons. These microscopes are capable of 200,000 magnification. X-ray microscopes work similarly and have the same magnification potentials as electron microscopes.

The specimens for transmission light microscopy are usually colored with dyes called histological stains. Stains help to identify the chemical makeup of the specimen and help visualize parts that are normally invisible. Light passing through the stain is altered by the dye and permits the scientist to see the specimen as a colored image. Dissection scopes do not need stain to visualize the specimen. Focusing of light microscopes is done by adjusting the distance between the specimen and various types of magnification objective lens.

Microtome

The instrument called a microtome gets its name from the scientific terms "micro" meaning small and "tome" meaning to cut. Microtomes are specialized instruments found in biotechnology laboratories that have a need to view cell structures under the microscope. These instruments are commonly used to prepare cells that are placed on light microscope slides. However, there are specific types of microtomes that prepare cells for viewing under electron microscopes. The earliest way of preparing microscope specimens was the sectioning of fresh or preserved samples by hand using a sharp knife or razor. Sections had to be thin enough to be almost transparent. This was a difficult task to do accurately and consistently. Various specimen holders called hand-held microtomes that make cutting easier developed over time. It took the development of the microtome for scientists to have specimens that provided the details needed for biotechnology research.

All microtomes have four major components: base, knife, knife attachment, and specimen holder. These four components can be modified into one of the eight modern types of microtomes. The base holds a hand-operated or motor-driven piston that the specimen holder moves up and down. This produces a cutting motion when the specimen is placed over a cutting surface. Specimens are preserved and mounted in a block of ice, wax, plastic-like materials, or resin. The microtome knife is mounted to a stationary knife attachment. Microtome knives are made of diamond, glass, sapphire, or stainless steel. Another device moves the specimen holder in a forward motion as the knife is cutting. The amount of forward motion is adjustable. This permits the scientist to make thick or very thin specimens for different types of biotechnology needs. Thick specimens are used for certain genetics studies, whereas very thin sections are valuable for looking at fine details of cells.

The most common type of modern microtome is the rotary microtome. Its base has a hand-operated or motor-driven rotating handle that moves the specimen holder up and down. It uses a very sharp stainless steel knife that cuts the sample which is mounted in paraffin wax. Rotary microtomes are good for cutting specimens as thin as 0.5 micron. Another traditional microtome is the sledge microtome. These are not usually found in biotechnology applications. They are more useful for cutting wood and other hard materials. The specimen holder is mounted on a sledge-like device located in tracks on the base. A forward and backward motion of the specimen brings it in contact with the steel knife. They can be mounted at a variety of angles to produce specific types of cuts. Biotechnology laboratories that produce wood and other fibers are likely to have these microtomes.

A freezing microtome uses a base similar to the rotary microtome with a special device that rapidly freezes the specimen on the specimen holder. These microtomes are unusual because the knife is moved up and down while the specimen holder is kept stationary. A fine jet of liquid carbon dioxide, liquid nitrogen, or a refrigerant is used to freeze the specimen. Many freezing microtomes are equipped with a similar device that keeps the steel knife cool. This prevents the specimen from melting due the friction caused by the cutting action. Freezing microtomes are not commonly found in biotechnology laboratories. They are mostly used for sectioning samples of food and textiles. Agricultural biotechnology facilities use them for food analysis.

One of the earliest microtomes was called the rocking microtome. They were invented in the early 1900s for cutting specimens embedded in paraffin wax. A hand crank or motor in the rocking microtome moves the specimen holder through an arc as it advances toward the knife. Rocking microtomes use a slightly curved knife called the Heiffor knife which is kept stationary. These microtomes are not good for making very thin specimens. They are usually used for tissue specimens studied in biotechnology laboratories. The cryostat is often confused with the freezing microtome. A modern cryostat is usually composed of a rocking microtome placed in a refrigerated housing. They are very common in animal agriculture and medical biotechnology laboratories. The chilled cutting conditions are useful for working with living animal and human tissue samples.

The saw microtome is highly specialized for slicing very hard specimens that are embedded in resins. As their name implies, the saw microtome uses a special saw blade to cut through the hard specimens. The specimen holder is moved slowly against a diamond coated saw that rotates at approximately 600 revolutions per minute. The blade is capable of slicing 20 μm or greater sections. Saw microtomes are found in biotechnology laboratories that work in bone and dental research and medical device materials production. The vibrating microtome is an uncommon device used in some biotechnology applications. It is primarily valuable for cutting fresh samples of soft tissue being studied for pathology or genetic investigations on animals and plants. A high speed motor vibrates a knife or razor blade attached to a knife holder. An electrical circuit adjusts the speed of vibration. The specimen is brought into contact with the knife and can be cut into various sized pieces depending on the speed of vibration.

Mixer

Mixing a solution of chemicals used in biotechnology applications is as critical as stirring a combination of foods for making a meal tasty. An unmixed solution may not carry out its intended jobs. Laboratory mixers are present in almost every biotechnology laboratory. They ensure that solutions or cultures of cells are homogeneous. Homogenous refers to the property of a mixture in which all the constituents are uniform throughout. Uniformity is particularly important for biotechnology products so that the drugs or items made from the products work consistently and according to their expected application. Even the simplest of biotechnology products such as milk is homogenized to keep it tasting the same from the first to the last sip.

There is no typical types of mixer. Different types of biotechnology mixing needs require different types of mixers. Mixers are categorized according to the type of mechanic action that does the mixing. The different types are propeller stirrers, bar stirrers, vortex mixers, shakers, tumblers, blenders, rotary mixers. Each of these mixers come in a variety of sizes. Small laboratory mixers can handle samples of chemicals or cells held in microliter containers. Large industrial mixers can mix thousands of liters of liquids and powders at a time. In some biotechnology operations the mixers are used outside in a location near large storage tanks that hold chemicals that must be mixed for bioprocessing operations.

Stirrers are generally used for dissolving chemicals in a solution or for ensuring homogeneity in a solution of chemicals or cells. Propeller stirrers are the most versatile means of mixing small to large amounts of solutions and cells. A typical propeller stirrer is composed of a motor, a motor control panel, and a mixing propeller. Most propeller stirrers use electrical motors. However, a majority of large industrial propeller stirrers use air-driven motors that operate in a manner similar to the power tools used in automobile repair shops. The motors are capable of being adjusted to spin at a range of 5 to 1,000 revolutions per minute.

Control panels for adjusting the motor speed can be a simple dial or a computer-operated unit that monitors the speed, temperature, and torque of the motor. Certain motors are called fixed-speed motors and operate only at one speed. These stirrers are used for one type of operation and are usually permanently attached, or dedicated, to a piece of equipment. They are commonly found in large bioprocessing facilities and are often attached to bioreactors. The propellers of stirrers vary greatly and must be carefully selected for their intended use. Propellers for cells and delicate chemicals such as DNA and proteins are designed not to destroy or shear the substances during mixing. A particular type of propeller will work better in specific kinds of containers of vessels.

Bar stirrers are usually small units composed of a magnetic stirrer and magnetic stirring bar. They are generally used for dissolving small to medium volumes of chemicals into solution. Special bar stirrers are used in small bioreactors. Almost every research laboratory has one or more bar stirrers because these stirrers are simple to use and are inexpensive compared to propeller stirrers. A typical magnetic stirrer is made up of an adjustable electric motor that spins a disk-shaped magnet. The motor and magnet sit underneath a holding platform. Many types of bar stirrers are available for different types of stirring needs. The bar stirrer is placed in a container in the solution that needs to be mixed. It is important that the bar is situated just above the spinning magnet to ensure consistent stirring.

Vortex mixers are another common piece of equipment in biotechnology laboratories. They are usually used for rapidly mixing small volumes of solutions in test tubes or other minute containers. A typical vortex mixer is composed of a vibrating motor and a sample holder. The vibrating motor moves the sample holder back and forth so that the solution in the tube or container spins. This motion produces a vortex or whirlpool in the solution. A vortex is a powerful circular current of water that facilitates mixing. Most vortex mixers have variable speed motors that can be adjusted for gentle or vigorous mixing. Some are designed with a touch switch that turns on the motor automatically when a test tube or container is pressed on the sample holder. Specialized vortex mixers can hold multiple samples.

Shakers or platform mixers mix samples in a similar manner as vortex mixers. They differ from vortex mixers because they have platforms that can mix large volumes of liquids. Shakers are composed of a variable speed motor attached to a platform that holds containers or racks for holding the sample. The shaker motor produces a rocking motion that stirs the contents of the containers. Microbial cultures are sometimes stirred slowly for days using shakers. A tumbler is generally used to mix thick solutions or dry powders. They are composed of a container attached to a belt-driven motor that rotates the container. Tumblers are often used in outdoor operations to mix drums full of chemicals in agricultural and environmental biotechnology applications. Large bioprocessing operations use vibrating tumblers or sifters that are used to separate different components of powdered chemicals. These tumblers use a porous vibrating platform to sift out certain sized particles. Sifters are commonly used in food industries and pharmaceutical manufacturing operations to ensure uniformity in their products.

Blenders are industrial mixers that are usually used for combining powders into a uniform composition. They are composed of a powerful motor that spins a blending paddle housed in a chamber. The paddles are designed for different types of materials ranging from thick liquids to coarse powders. Chambers can be designed to hold a particular volume of material or can be set up to mix a continuously flowing supply of material. Blenders are usually monitored with computers to ensure uniform operation. Rotary mixers work much like a sideways washing machine. They are composed of a rotating drum powered by a motor housed in a casing. Rotary mixers are commonly used in large-scale biotechnology production where they are sometimes called batch mixers. Paddles lining the inside of the drum can be designed to rapidly blend batches of ingredients to uniformity without causing sensitive chemicals to clump, degrade, or smear.

Nanotechnology

Nanotechnology is a broad category of biotechnology that uses microscopic machines to carry out a variety of tasks. The size scale of nanotechnology is incredibly small. Most of the nanotechnology applications are no larger than one four thousandth of an inch in size. Biological molecules can be molded into miniature "gears" formed from individual atoms. The term "nano" comes from the metric unit nanometer which is a billionth of a meter. Nanotechnology was named by K. Eric Drexler in 1986 in his book about the future of mechanical technologies called *Engines of Creation*. One field of nanotechnology called bionanotechnology makes use of biological molecules to create instruments that replace the functions of traditional electronic circuits and mechanical devices. It is a rapidly growing area of biotechnology being developed for a variety of medical applications.

Bionanotechnology developments include nanoparticles being investigated as a drug delivery system. Nanoparticles can carry small amounts of drugs to specific cells unlike traditional medicines that can enter every body cell. One was recently produced to carry anticancer drugs to tumor cells. Bionanomachines are devices using carbohydrates, DNA, and proteins in place of traditionally used metals and plastics. The molecules can be molded and moved into microscopic lightweight robots. DNA computers use DNA molecules to store bytes of information with better efficiency than standard silicon computer chips. Biosensors can be injected into the body as a means of monitoring a person's blood chemistry. Other bionanotechnology inventions are being developed for a variety of purposes including the detection of pollution to the removal of clots from clogged arteries.

Nuclear Magnetic Resonance Imaging Instrument

Nuclear magnetic resonance imaging (NMR) or magnetic resonance imaging (MRI) is an analytical technique used to study the chemistry of animals, plants, and biotechnology products. NMR is capable carrying out many types of analytical studies useful for biotechnology. It is an important way of characterizing the chemical structure of purified molecules produced by bioprocessing. This is a very important application in the production of diagnostic and pharmaceutical chemicals. NMR is useful in determining the shape and structure of DNA, proteins, and other complex molecules used in biotechnology applications. Researchers who study the metabolism of genetically modified cells are also finding NMR valuable for many cell analysis applications. Whole living animals and plants used in biotechnology applications can also be studied using MRI instruments similar to those used in medicine.

Most NMR instruments are composed of two magnets, sweep coils, sample holder, radio frequency transmitter, amplifier, readout, and control panel. The NMR's magnet is the most noticeable component of the instrument. Two large magnets are oriented to produce a tremendous magnetic field and average 1 cubic meter in size. They are housed in a large insulated jacket that must be cooled with one of the few types of refrigerants called cryogens that circulate through the jacket. Liquid helium is the most common refrigerant. An electrical circuit called a cryostat maintains the temperature of the magnet and reduces helium evaporation rate. The cooling of the magnets must be monitored carefully to keep them from breaking.

Lining the jacket is a shield that prevents the electric field from passing through the jacket. The electric field is strong enough to damage other electrical equipment and would even be harmful to people. The size of the magnetic field is proportionate to the volume of the sample being studied. A 1 cubic centimeter sample requires 1 cubic meter of magnets. Industrial NMR instruments use large magnets in housings taller than a house. Miniature NMR instruments are being developed to handle the minute samples frequently encountered in biotechnology research laboratories. A vibration reduction system is also built into the magnet housing to dampen any ground vibrations that may affect the performance of the instrument.

The sample holder of the NMR is a tube attached to a rotating motor. Surrounding the sample holder is a radio frequency transmitter. The sample holder and radio frequency transmitter are positioned in the magnet housing so that they are situated in a strong magnetic field. Two sweep coils are positioned between the magnets and the sample holder. Sweep coils are one or more loops of an electrical conductor used to create a radiofrequency. These are sometimes called radiofrequency coils. The role of the sweep coils is to produce an electrical field that detects changes in the sample as the chemicals in the sample are exposed to the magnetic field. These changes are detected by the radiofrequency transmitter and sent to the amplifier. The amplifier then feeds into a computer that converts the data into information about the sample.

NMR provides information about the position of specific atoms within a molecule. It does this by using the magnetic properties of an atom's nucleus which is typically composed of protons and neutrons. Some researchers categorize NMR as a form of spectroscopy because it measures the absorption and emission of energy arising from changes in atoms exposed to electric field. Spectroscopy is usually defined as the measure of a chemical's absorption or emission of light energy. NMR detects the positions of an atom by measuring a feature of the atom's nucleus called the spin state. Spin state is based on the principle that nuclei of all elements carry characteristic charge. The spin state is determined by the number of protons and neutrons making up the nucleus. Each different element has a unique number of protons. If the number of neutrons and the number of protons are both even, then the nucleus has no spin. The atoms have a half-integer spin when the number of neutrons plus the number of protons is odd. The amount of half-integer spin, for example 1/2 versus 3/2, is proportionate to the complexity of the nucleus. If the number of neutrons and the number of protons are both odd, then the nucleus has a full integer spin. The spin properties of elements are compared to that of hydrogen which is the simplest of the elements.

The role of the magnetic field is to make it possible to measure the electrical force given off by the atomic nucleus's spin state orientations. With no electric field the spin state orientations are of equal energy and thus not capable of being measured. The magnetic field causes the atoms to stay at one or another of its energy levels. This is called an energy level split. Each level is given a magnetic quantum number value that is characteristic for an atom. The magnetic quantum number is a part of the science of quantum mechanics. Each atom has a particular magnetic quantum number based on its theoretic number of electrons in the atomic orbitals or shells. The number of electrons of a pure element is the same as the number of protons. NMR provides a picture of the number of protons. The strength of NMR is that it can produce a three-dimensional image of a biotechnology chemical showing the location of each atom.

Particle Sizer

As evident in its name, a particle sizer or particle size analyzer measures the size of large chemicals and whole drugs produced in biotechnology processes. These instruments are most commonly found in pharmaceutical manufacturing facilities. They provide an accurate measurement of particle size that is important for determining the consistency, quality, and performance of a biotechnology or pharmaceutical product. Government regulations in various countries have quality control guidelines, such as the United States current Good Manufacturing Practices regulations (cGMP), and set criteria for consistency of particle size for many biotechnology products. The particle sizer that is selected for a particle sizing application must be appropriate for the type of material being measured. It addition, it must be designed for the conditions it will be used in. For example, certain instruments are designed for laboratories while others are used in industrial situations. The data provided by the particle sizer must meet the specific needs of the biotechnology application.

Most modern particle sizers use optical devices to measure particle characteristics. This method is able to determine the size of bubbles, droplets, or solid particles of various sizes. Particle sizers are designed for specific size ranges between 130 millimeters to 0.1 micrometers. One type of particle sizer is called an aerosol analyzer and it converts the sample being tested into a liquid or spray. Solid particle analyzers or screener and related instruments called shape analyzers measure solid chemicals and pills. A typical particle sizer is composed of a light or energy source, particle handler, detector, and readout. Particle data is usually recorded as irregular, oversize, or undersize. Particle sizers can be used a stand-alone instrument or can be incorporated into machinery to give real-time information during manufacturing or bioprocessing operations.

The aerosol analyzer uses a principle called particle dynamics analysis to measure the size and speed of particles. Samples are placed in the particle handler that uses a spray head to convert the particles into an aerosol. The spray passes through a window where it is exposed to a beam of light from a bulb or a laser. Particles in the spray interfere with the beam of light and scatter the light based on their size and shape. The speed of the particle is determined by measuring how fast the particle traveled through a series of beams. Aerosol analyzers have a series of detectors that collect the scattered light for each particle passing through the beam. A computer linked to the detector then uses mathematical formulae to determine the size, shape, and speed of the particle. Particle speed is important because it provides information about the weight of the particular chemical making up the particle. The scientist then gets a readout on a computer screen or on a printed graph that shows the characteristics of the particles in the sample. These analyzers are commonly found in many types of biotechnology research laboratories and industrial facilities.

Particle screeners have a different type of particle handler than aerosol analyzers. The samples are placed through a size sorter that is made up of one or more screens that separate the particles based on size and shape. They are many types of size sorters. Some particular screeners separate the particle using a rotating drum. Others use a vibrating screen that forces particles through select openings in the screen. The particles then travel past a light source. Each particle blocks and scatters the light as it passes through the beam. A group of detectors measure the number, size, and shape of the particles. Velocity can also be measured. The detectors feed the information to a computer that converts the data into useful statistical analyses. A readout is provided as an image on a computer screen or as a printed graph. These analyzers are typical of pharmaceutical companies.

pH Meter

The scientific measurement called pH is essential for any chemical mixture needed for biotechnology procedures. Cells grown in culture must have the proper pH to keep the cells alive and healthy. Chemical reactions for in biotechnology procedures must have the precise pH for the reactions to take place. Moreover, a certain pH is needed to permit the function of enzymes used in the reaction. pH is a mathematical value that represents the concentration of hydrogen ions in a solution. The value is placed on a scale called the pH scale that normally ranges from 0 to 14. This scale reflects the concentration of hydrogen ions in a solution. The lower numbers denote acidic conditions which are defined by having a large concentration of hydrogen ions in solution. Higher numbers represent basic, or alkaline, conditions. This is typical of solutions with low amounts of hydrogen ions in solution. A condition called neutral is signified by the number seven. Each decrease in the value on the pH scale represents a ten-fold increase in hydrogen ion concentration. Therefore, a pH 6 solution has 10 times more hydrogen ions than an equivalent volume of solution at pH 7.

The concept of acid and base goes back to the ancient Greek scientists. However, the pH scale was developed by the Danish biochemist Soren Sorensen. He invented a device similar to the modern pH meter to come up with the pH scale. A pH meter is an instrument that detects the hydrogen ion concentration of a solution. In effect, a pH meter determines the hydrogen ion concentration by measuring the ability of an electrical current to pass through the solution. Hydrogen ions have a positive electrical charge and can pass an electrical current. Current travels better through the solution as the hydrogen ion content increases. A typical pH meter is composed of an amplifier, a readout, and an electrode. The amplifier magnifies the current so the readout is able to function. Readouts can vary from a digital readout that displays the pH number to an analog readout in which a needle points to a number on a printed pH scale.

The pH electrode measures the electricity conducted by the solution. It is divided into a measurement component or glass electrode and a reference component or reference electrode. The glass electrode is a special glass tube containing concentrated salt solution and a piece of metal used to detect electrical current. Hydrogen ions from the solution being tested pass through microscopic passages in the glass tube and change the electrical properties of the salt solution in the tube. Acidic solutions transfer more hydrogen ions into the glass electrode than would a basic solution. The electrical conduction measured in the glass electrode is then compared to a current in the reference electrode. A typical reference electrode contains a known amount of acid and is also attached to a metal wire that measures current. It produces a particular current for a known amount of pH determined by the acid in the electrode. Both currents pass through the amplifier and into a circuit that compares the glass electrode's current to the current in the reference electrode. This circuit then calculates the pH and then passes the information along to the readout.

The pH electrode is very sensitive to the effects of temperature. A pH meter compensates for this by mathematically adjusting its measurement of the electrical current to the temperature of the solution. Before pH electrodes were invented scientists relied on simpler, but less accurate, ways of determining pH. Certain chemicals change their color in response to hydrogen ion concentration. They are collectively called pH indicators. Certain pH indicators change color in response to acids, while others change color in response to bases. The first chemical discovered to have these properties was a substance extracted from an organism called lichen. This substance was called litmus, being named for the scientific designation for the particular lichen. Scientists dissolved these pH detecting chemicals onto a sheet of paper and then dipped the paper into a solution to determine the pH. Mixtures of pH indicators can be formulated to detect a wide range of pH values. Many scientists still use pH indicators when they need to get a rough estimate of a solution's pH.

Pipette

Pipettes are a fundamental tool found in every biotechnology laboratory worldwide. These workhorses of the laboratory are precision instruments for measuring and delivering precise amounts of liquids. The simplest pipette is a narrow glass or plastic tube marked with specific liquid volumes. These large pipettes are controlled manually and are used for delivering volumes ranging from 0.1 to 10 milliliters. Autopipettes or micropipettes are designed to accurately and consistently transfer volumes as low as hundredths of a milliliter. They are controlled by a plunger that is set to deliver a set amount of liquid. The plunger can be operated by hand or by robotic attachments. Micropipettes use small plastic tubes called tips that collect the liquid. The fluids being transferred never enter the component containing the plunger. Some micropipettes can hold several tips allowing the simultaneous delivery of samples. These tips are thrown away after each use to prevent contamination of experiments and to ensure any hazardous chemicals being transferred are properly disposed.

A typical pipette is used in the following manner. First, the user must dial or program in the volume that needs to be transferred. A tip is added to the pipette's delivery system. The plunger on the delivery system is then pushed down to a set point as the tip is placed in the liquid sample. Releasing the plunger slowly then siphons up some of the preset amount of sample. The pipette is then positioned so the tip is placed over the location when the sample is being transferred. Another press on the plunger releases the sample. Some pipettes require a second plunger action to remove any residue sample remaining in the tip. The tip is then discarded and replaced with another tip before using the pipette again. For most pipettes to perform accurately and consistently, some training and practice is required.

Polarimeter

Polarimeters or optical rotation instruments measure a property of certain biological molecules called chirality. Chirality is a critical aspect of many biotechnology products. It is most easily defined as the chemical equivalent of left and right hand. To a scientist interested in chemical analysis, chirality means the ability of a biological chemical to exist in one or two mirror image forms or enantiomers. Each enantiomer rotates polarized light in opposite directions. An enantiomer that rotates light in the right direction, or clockwise, when viewed head-on looking at a beam of light is called the dextrorotatory (d) or (+) enantiomer. The levorotatory (l) or (−) enantiomer bends light to the left. Polarized light refers to a beam of light whose waves are uniformly aligned in one direction or travels in a single plane.

Biological molecules that are made by organisms are usually one type of enantiomer form. All of the basic building blocks of carbohydrates are produced in the dextrorotatory form. The amino acids used to build proteins are made in the levorotatory form. Every organism on Earth has a metabolism that is limited to using dextrorotatory carbohydrates and levorotary amino acids. The other form can be toxic or lethal if taken in large amounts. Enantiomer forms differ in their biological activity in the body. Biological activity refers to the ability of a molecule to cause metabolic changes in a cell. For example, the dextrorotatory form of glucose tastes sweet and can be used by the body to produce energy. The levorotatory form is tasteless and blocks metabolism causing illness.

The optical rotation properties of biotechnology products are very important for determining the effectiveness and safety of the product's chemistry. Strict governmental guidelines require that biotechnology products used in medicine are composed of a pure form of the compound. In addition, many drugs that are synthesized by scientists are regularly contaminated with the inappropriate enantiomer. This makes it important to detect and remove the undesirable enantiomers. Scientists are also learning that the opposite enantiomer forms are the basis for useful therapeutic agents. For example, the levorotary form of glucose is being tested for inhibiting cancer cell growth.

Polarimetry of enantiomers is based on the principle that a change in direction or the rotation of a plane of polarized light occurs when the light is passed through an optically active substance. Enantiomers are optically active substances. A sample that contains only one enantiomer of a chiral molecule is said to be optically pure. Optical purity is assigned a measurement based on the percentage of light that bends in a particular direction. A polarimeter contains a sample chamber, light source, polarizers, detector, and readout.

A beam of light from the light source passes through the polarizer. The polarizer filters the light so that the light waves are all traveling in one plane that can divert either right or left when it encounters an optically active substance. This polarized beam of light is then passed through the sample that is dissolved in a particular solvent and is enclosed in the sealed sample chamber. The light then hits the detector which measures the amount of light hitting it at particular angles to the original plane of light. A computer then calculates the optical purity of the sample and sends the data to a computer readout or a graph paper output.

Rheometer

Rheometers are specialized instruments that measure a property of liquids and polymers called rheology. Rheology determines the ability of a material to flow or be deformed. The term was named in 1920 by Professor Eugene Bingham at Lehigh University in Pennsylvania. A material's ability to flow is called viscosity. Viscosity is defined as the thickness or resistance to flow of a liquid. Materials with a high viscosity do not flow readily, while materials with a low viscosity are more fluid. For example, syrup would have a high viscosity and water would have a low viscosity. Viscosity is sometimes measured in biotechnology manufacturing processes. It is an important indicator of the concentration of a particular compound in solution. Many biological molecules thicken the solution they are dissolved in as their concentration increases. It is also important to know the viscosity of a solution that is being mixed in large vats. Many of the motors and pumps used in mixing must be adjusted to match the viscosity of the solution to keep from burning out or malfunctioning.

Rheology also provides insight into the molecular structure of polymers used in biotechnology applications. Using a measure called elasticity, rheology can determine various structural details about the polymers including branching patterns, shape, and size. Elasticity is defined as a polymer's tendency to return to its original shape after it has been compressed, bent, stretched, or twisted. A variety of novel carbohydrates and proteins produced by biotechnology methods are characterized this way to help identify uses for the products. Rheology can also be used to tell various chemical properties of polymers such as its degree of decomposition, response to temperature, and stability. Solutions of DNA or RNA can be analyzed with rheometers to make sure that they are the proper size and purity. The consistency of large batches of biotechnology products is often determined using rheology. Consistency is especially important in the production of pharmaceutical compounds. It is a crude measure of the dose of the drug in certain volume of fluid being processed into a drug. Cheese manufacturers even use rheology to tell the elasticity and uniformity of different types of cheeses.

Rheology can be carried out using either a capillary, dynamic rotational, or torque rheometer. A capillary rheometer consists of a heated barrel, a piston, and a small chamber. The piston moves the material being tested through a very small opening called the capillary into the chamber. A typical opening is 0.75 millimeter in diameter. This type of rheometer heats the material and then monitors the way the material reacts to certain conditions in the chamber. The person operating the capillary rheometer can change the shape of the chamber and speed of the piston to analyze different rheological properties. A dynamic rotational rheometer is composed of a motor, a holder for applying force on a material, and a torque-sensing mechanism. These rheometers are mostly used for studying thick polymers. The motor causes the holder to crush, stretch, or twist the polymer while the torque-sensing mechanism sends a signal to a computer. The computer then provides rheological information about the chemical. Torque rheometers resemble small mixers. Its major component is a special motor that can measure the torque or twisting action of a solid polymer or a solution. The mixer is attached to a sensor that then feeds information into a computer. It can measure the viscosity, stretching, and twisting of a material.

Spectrophotometer

Spectrophotometers are very likely found in every biotechnology laboratory. They are important instruments for identifying various biological molecules. These instruments are also useful for determining the concentration and purity of almost all biological molecules. Many biotechnology applications also use it to monitor chemical reactions carried out in bioprocessing activities. Spectrophotometers can be used alone or can be integrated into other laboratory instruments. Some bioprocessing equipment have built-in spectrophotomers that monitor the cell or chemical processes used in manufacturing biotechnology products. The term spectroscopy refers to the observation (scopy) of various wavelengths of light (spectro). Spectroscopy uses light to identify and determine the concentration of a particular chemical in a solution. It makes use of the ultraviolet, visible, and infrared regions of the light spectrum. Humans perceive different types of light as color. However, scientists measure light as wavelength. Wavelength is defined as the distance between repeating points of a wave of light. Red light has a long wavelength while blue light has a short wavelength.

There are many types of spectrophotometers used in biotechnology to measure a variety of chemical properties. The basic spectrophotometer is composed of a light source, monochromator, beam path, sample holder, photodetector, and readout. Most spectroscopy uses a light source made up of a lamp that produces visible light very much like a household light bulb. Most spectroscopy is done using a visible light lamp. These lamps provide white light which is made up of the whole

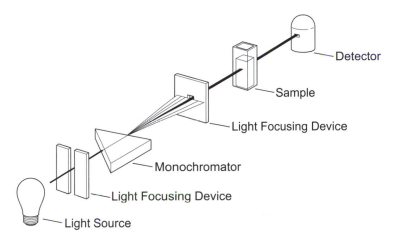

Figure 3.9 A spectrophotometer is used to analyze chemicals in a solution by passing a beam of light through the sample. (*Jeff Dixon*)

visible light spectrum. Infrared or ultraviolet light lamps can also be used as a light source. The type of light used depends on the nature of the chemical testing being performed. Lamps usually have to be warmed up and maintained at a certain temperature to ensure consistent lighting during each use. An improperly operating lamp can give the wrong information without the operator being aware of any errors.

All spectrophotometers must be able to provide a pure beam of light consisting of a few wavelengths that is then transmitted to the chemical sample. A pure beam of light is essential for ensuring accurate and precise chemical measurements. This pure beam of light is produced by using a device called a monochromator. A monochromator is an adjustable crystal, filter, or mirror that isolates portions of the light spectrum by separating the light into its component wavelengths. Most chemical measurements use a narrow range of wavelengths. The range of wavelengths provided by a monochromator is called the bandwidth. Many modern monochromators use an adjustable control called a slit, or collimator, to control the bandwidth. The slit focuses the light through a series of lenses called the beam path. These lenses pass the beam along to the sample. Certain spectrometers are designed with a monochromator that splits the light into two beams. These dual-beam spectrophotometers are able to measure and compare two samples at a time. They are very useful for investigating the chemical purity of many biotechnology products.

The sample holder of the spectrophotometer is sometimes known as the cell or well. It is an opening in the spectrophotometer that places the sample between the beam path and the photodetector. The sample holder of most spectrophotometers contains a clip that grips a special container called a cuvette. A cuvette is a container that is sealed at one end and holds a precise volume of the sample being studied in the spectrophotometer. Cuvettes can be cylindrical or square and are made of glass, plastic, quartz. The type of cuvette being used is determined by the wavelength of the beam. Ultraviolet light spectroscopy requires quartz cuvettes because glass and plastic interfere with the passage of ultraviolet light to the sample. Some sample holders can carry several cuvettes that are then passed into the beam path one at a time. Certain spectrophotometers use specialized holders for analyzing flowing samples and solid materials.

The photodetector, as the name implies, measures light. In absorption spectroscopy the photodetector measures the amount of light that has passed through the sample. This measure is called absorptance or transmittance. Absorptance measures the amount of light that is absorbed by the sample. Transmittance measures the amount the light that has passed through the sample. Another type of photodetector measures the glow given off by a sample that is exposed to a particular wavelength of light. Fluorescent spectrophotometers use this type of photodetector. The photodetector converts the light into an electrical signal that is displayed as a measurement on the readout. The readout can be a digital readout that displays the measurement as a number to an analog readout in which a needle points to a number on a printed scale. Many readouts are associated with a computer that can modify the readout into other measurements. Fourier transform spectroscopy uses a computer that converts the readout in a graph containing a series of complex curves. The curves act like a fingerprint for a particular molecule or mixture of molecules.

Spectrophotometry can be specialized to measure the mass of a molecule using a technique called mass spectroscopy. In mass spectroscopy the sample is heated and electrically charged to form a gas and is then passed through a special series of magnets that move the sample through the tube at a very high speed. Another magnet separates the components of the molecules. The components are then directed to a detector that measures the impact and electric charge of the various components. A computer then interprets the information from the detector for the size and composition of the molecule. Mass spectroscopy is a very important tool of providing the characteristics of molecules whose structure is not currently known. It is also used to distinguish different molecules such as many types of drugs that have very similar structural properties.

Another specialized type of spectroscopy is atomic absorption (AA) spectroscopy. It is used to identify charged particles called ions. They are most useful for indicating the presence of metals and salts in a sample. Many drugs must contain a particular type of salt in order to be used by the body. In addition, certain metals are harmful to humans and can poison organisms used in biotechnology processes. Atomic absorption spectroscopy uses the absorption of light to measure the concentration of ions. The sample is usually a liquid or a solid that is analyzed by first vaporizing it in a flame or furnace. As the ions are heated they absorb particular wavelengths of ultraviolet or visible light specific to the type of ion. The ion concentration is determined by the amount of the light absorbed. Absorptance is measured using a photodetector that sends the signal to a readout that displays a pattern called an absorption spectrum. The pattern of the absorption spectrum is then interpreted as a particular ion by the scientist or by a built-in computer.

Thermocycler

A thermocycler is laboratory instrument that repeatedly cycles through a series of temperature changes required for chemical reactions such as the polymerase chain reaction or PCR. The polymerase chain reaction is a technique used to make multiple copies of DNA fragments. This process is called amplification because it can generate hundreds to thousands of copies of DNA. PCR is used to produce ample of quantities of DNA when only a small amount is available. It is a valuable tool for DNA analysis, disease diagnosis, and genetic engineering. The technology is regularly used in crime scene analysis to collect DNA from traces of blood, hair, saliva, or skin.

The effectiveness of a thermocycler is its ability to change temperature rapidly and with precision. Each PCR reaction has an exacting set of conditions needed to get accurate copies of the desired DNA. Part of the heating and cooling efficiency is due to the small sample containers used in the thermocycler. Samples of DNA to be copied are placed in minuscule containers called microtubes. For PCR it is not unusual to work with volumes of solutions less than 5 hundredths of a milliliter. The DNA fragment that is copied is placed in the microtubes containing a special mixture of solution that helps build new copies of DNA. The researcher programs the thermocycler for a series of three-stage cycles. Stage one is a hot stage called denaturation. The approximately

90°C temperature opens up the DNA for copying. This is followed by a cooling or annealing stage that permits the DNA to attach to chemicals needed to copy it. Stage three is a warm temperature cycle called the elongation cycle. It encourages the growth of the DNA copies.

The thermocycler is set to run a particular number of these cycles depending on the amount of DNA a person wants to collect. Thousands of copies of DNA can be made this way. PCR was carried out before thermocyclers were invented. The procedure was a laborious and time-consuming activity that involved placing the samples in container full of heated water or ice. It was very difficult to get consistent results using the older method. Moreover, a person could spend hours nonstop in the laboratory just doing a PCR procedure.

Thermometer Probes

Thermometer probes or digital thermometers have replaced the traditional alcohol or mercury thermometers using in scientific applications. The tradition thermometer measures the contraction or expansion of liquid alcohol or mercury in a sealed tube. Most liquids expand or contract in response to temperature. Temperature can be simply defined as the measure of how fast the atoms and molecules of a substance are moving. Slowly moving atoms bounce off each other with very little force and thereby remain close to each other. This keeps the substance compressed. Cooling the substance slows the particles and permits them to come closer together causing the substance to contract. Rapidly moving atoms or molecules bounce against each rapidly and consequently move further apart. This movement causes expansion of the material. Traditional thermometers are calibrated so that the liquid contracts or expands to a numbered marking that indicates the temperature in Celsius or Fahrenheit degrees.

The traditional thermometer is not accurate and consistent enough to be used in most biotechnology applications that require precise temperatures. Contraction and expansion of the liquid can vary greatly from one thermometer to another based on a variety of factors that cannot be fully controlled while manufacturing the thermometers. In addition, they do not readily respond to rapid temperature changes that must be carefully monitored in many biotechnology processes. It is also very difficult to design tiny easy-to-read traditional thermometers that can be used to check the temperature of minute samples. Traditional thermometers are also easily broken and are not effective in very cold or tremendously hot temperatures. The liquids can freeze or vaporize respectively under those conditions. Scientists began using digital thermometers in biotechnology operations in response to the limitations of traditional thermometers.

Digital thermometers can be designed for accuracy and consistency under a variety of conditions. They are capable of measuring extremely cold or hot temperatures. In addition, the sensor used to detect the temperature can be made so small that they must be viewed with a microscope. Digital thermometers have been designed to measure temperatures as low as -270°C and as high as 3000°C. One type of digital thermometer uses a device called a thermocouple to detect temperature. A thermocouple is formed by combining two strips of dissimilar metals side-by-side in a sheath called the probe. The junction between the two metals produces a small bend in response to the temperature. This bend is caused because the metals expand and contract at different rates. The one that expands more quickly bends back on the other causing the metals to curve. This bending is then attached to an electrical circuit that converts the bending into an electrical charge. The circuit is attached to a readout that converts the electrical energy into a calibrated temperature reading in Celsius, Fahrenheit, or Kelvin units.

Another type of digital thermometer uses a resistance probe to measure temperature. It looks like the thermocouple thermometer but measures the temperature using a different principle of physics. The simplest resistance probe is formed by a metallic conductor connected in an electrical circuit that forms a closed loop. Resistance probes measure the resistance changes in the metals in the circuit. The resistance change is determined by a small circuit that compares the resistance difference of the metal as a small electrical charge is passed through the metals. Each type of metal has a unique range of resistance changes at various temperatures. Many resistance thermometers use platinum wire in the loop because it is a very good conductor of electricity. It changes resistance readily to electricity in response to small temperature changes. The resistance change is fed into a circuit attached to a readout that converts the electrical energy into a calibrated temperature reading in Celsius, Fahrenheit, or Kelvin units. Resistance thermometers have what is called a reference circuit built in to the system to ensure accurate measures each time they are used.

Water Bath

As the name implies, a water bath is a tub of water used to bathe a chemical reaction or culture of organisms used in industrial and laboratory procedures. The primary job of the water bath is to maintain the chemical reactions or organisms at a constant temperature. Early scientists built water baths by placing a glass beaker into a container of water that was chilled with ice or heated with a flame to the desired temperature. Many scientists still use this type of technique today when they need a temporary means of storing small amounts of sample at a particular temperature. They still use ice as a cooling agent, but the flame has been replaced with an electric heater. However, this type of setup is not reliable for larger samples and cannot maintain constant conditions for long-term temperature control.

Electric water baths are the preferred way of temperature maintenance for procedures carried out in almost all biotechnology laboratories. The typical water bath is composed of tub, control panel, and temperature control unit. Most water baths are designed to warm the water from temperatures ranging from 20°C to 75°C. Cold water baths are also commonly used and are usually operated between 0°C to 15°C. Most chemical reactions and human cells used in biotechnology are maintained at 37°C, which represents the average temperature of the human body. The tub is usually composed of a stainless steel pan. Pan sizes vary in size from 500ml to 40 liters capacity. Samples are usually placed in beakers, flasks, or test tubes that are placed in racks immersed in the water-filled pan. Many water baths come with interchangeable racks.

The temperature unit is usually a refrigerator for cold water baths and an electric heater for warm water baths. Refrigerator units use coils filled with refrigerant or ethylene glycol to cool the water in the tub. The coils can be located within the tub or can line the outside of the tube hidden away from view. A pump continuously moves the refrigerant or ethylene glycol through the coils to maintain the temperature of the water in the tub. Most warm water baths have electrical heating coils located inside or outside of the tub. The water bath's control panel contains a dial or touchpad that allows the user to adjust the temperature. Some water baths permit the user to set heating and cooling cycles for special types of reactions that need variable temperatures.

Specialized water baths are used for particular applications needing other reaction conditions or growing environments. Many biotechnology laboratories have shaker water baths. These water baths have a movable tub that is rocked back and forth by a motor that can be adjusted to various speeds. Shaker water baths are needed for reactions or cell cultures that must be regularly mixed. Certain bioprocessing reactions require a boiling water bath. These baths are heated with steam coils that heat the water in the tub from 90°C to 100°C. The heat is used to deactivate certain chemicals in the reaction without affecting other chemicals. Moreover, some processes require that cells are "shocked" with heat in order to stress the cells into producing certain chemicals. Large enclosed water baths are used in many sterile bioprocessing operations. Racks are placed inside a large sealed drum that can be cooled or heated. A similar setup is used for the pasteurization of milk products. Pasteurization is the rapid heating and cooling of sensitive fluids such as milk and certain pharmaceutical compounds to kill certain types of bacteria.

Water Titrator

The purity of many biotechnology-derived products is determined by measuring the water content of the substance. This is particularly important for chemicals used in pharmaceutical applications that must have a specific amount of the active chemical in a certain mass of the product. Water tends to stick to many molecular mixtures and contributes to the mass of the substance. This has a dilution effect on the chemicals making them less likely to carry out their job at the expected levels. In addition, residual water can degrade the active components of many biotechnology products and affects many means of preserving the materials. Scientists have developed various ways to measure the presence of water in chemicals in response to the problems created by water.

The earliest method used to determine water content involved a precise dehydration process. Chemicals were weighed before and after being placed in a special oven used for water content analysis. The difference in weight was calculated as water percentage. This was not a very accurate method and did not take into account other substances that could have evaporated from the chemical. Water titrators are a quick and accurate way to measure water content. In addition, they provide consistent measurements that can be calibrated to match governmental product quality regulations. The Karl Fischer titration method is currently used in contemporary water titrators used in biotechnology laboratories. It is a widely used analytical method for quantifying water content.

The Karl Fischer method is based on the Bunsen reaction between iodine and sulfur dioxide in an alkaline alcohol solution. In this reaction, the alcohol reacts with sulfur dioxide and an alkaline substance to form a substance called alkylsulfite. Pyridine and imidazole are the commonly used alkaline substances for carrying out the Bunsen reaction. The iodine then converts the alkylsulfite to alkyl sulfate by a chemical reaction called oxidation. This oxidation reaction requires water in order to take place. A typical water titrator is composed of a reaction chamber, sample port, electrode, and readout.

A weighed sample is added to the sample port. The sample then drops into the reaction chamber containing the alkaline substance, iodine, and sulfur dioxide dissolved in alcohol. The presence of water in the sample drives the Bunsen reaction. Water concentration in the sample is calculated based on the concentration of iodine that was used to carry out the Bunsen reaction. The electrode measures the decrease in iodine as the reaction progresses. A sample that is high in water causes a greater loss of iodine than drier samples. An electrical signal from the electrode feeds data into a computer which then represents water content as percent water per gram of material. This information is displayed on the readout.

4

BIOTECHNOLOGY INNOVATIONS

THE CREATION OF INNOVATIONS

People are not necessarily born to be scientists. Scientists come from a variety of cultural backgrounds. They also have a wide array of family and religious upbringings. In addition, it does not always take an early interest in science to become a prominent scientist. Some people knew they wanted to be scientists as a young child. Others did not develop a concern for science until they entered college. Many scientists were not discouraged from achieving greatness in spite of the barriers and prejudices intended to exclude them from success. Some scientists discovered great things when they were in their twenties. Others did not come upon their fame until their fifties. There is no such thing as a typical scientist. Each scientist has his/her unique habits, hobbies, and lifestyles. However, one thing that all scientists have in common is a curiosity about the way nature works. This natural curiosity is fostered by an attitude to produce innovations. The drive to innovate science is very similar to the desire of others who contribute innovative ideas to architecture, art, literature, music, and poetry.

People who work in biotechnology usually focus on one fundamental area central to the principles of biology: cell theory, evolution by natural selection, gene theory, and homeostasis. Cell theory takes into account the way a cell functions. Many biotechnology applications include manipulating a cell's function in order to cure a disease. An understanding of evolution by natural selection is essential for producing new types of biotechnology organisms with characteristics that assist with their survival. Many new crops are developed with genes that protect them from diseases and harmful environmental changes. Gene theory explains the role of DNA in controlling a cell's functions including the passing on of traits to the next generation. Genetic engineering is a biotechnology application of gene theory that alters a cell's DNA in order to cure diseases or to control the metabolism of a cell for a specific purpose. Homeostasis must be taken into account with all biotechnology research to ensure that biotechnology applications function properly and pose few, if any, hazards to people and nature.

All scientists who work in biotechnology are not necessarily trained in biology. Many of them have chemistry and physics backgrounds. These scientists are able to make their accomplishments because they relate the principles of their scientific disciplines to biotechnology applications. Most scientists today realize that it is the accurate use of the scientific method that is most important in carrying out biotechnology research and development. Many philosophers and scientists pondered over the question "what makes a great scientist?"

Most great scientists would conclude that there is a single set of characteristics shared by great scientists. This is easily learned by reading the biographies of the various great scientists who contributed to biotechnology. Many of the great scientists were not interested in science when they were children. Not all of them did well in school. Probably the most important characteristic shared by most great scientists is a natural curiosity about how things work. They were also not afraid to rationally challenge current theories. Creativity, drive, motivation, and persistence are also important properties of great scientists. However, these characteristics had to be channeled in a direction that solved scientific problems.

Many scientists would claim that "luck" or "being in the right place at the right time" brought greatness to some scientists. This notion was downplayed by Louis Pasteur who stated that "luck favors the prepared mind." Albert Einstein and Isaac Newton also recognized that it was their creative thinking that took advantage of a lucky observation. In several instances, two scientists came upon the same discovery. However, the acclaim for the discovery went to the scientist who recognized the full significance of what he/she found. Some great scientists put themselves in lucky situations by seeking out to work with great scientists or to work in areas of science that had much promise. Michael Rosbash, a professor of biology at the Howard Hughes Medical Institute in Maryland, simply stated that to be a great scientist a person must "Follow your star. If you are interested in something, go for it."

HISTORY OF BIOTECHNOLOGY INNOVATIONS

Many people today mistakenly believe that biotechnology is a new science. It is also common for many people to narrowly interpret the scope of biotechnology as being genetic engineering. However, biotechnology is an ancient practice that even predates formal applications of science and technology. Much of early biotechnology was solely based on human observations of nature. The scientific explanations for early biotechnology applications did not come about until the 1940s with the discovery of metabolic pathways. However, the research leading to the study of cell chemistry required the first investigations into cell structure beginning in the 1830s. The modern applications of biotechnology that appear regularly in the news did not get started until the 1970s. Other fields of science have had the growth in information and technology exhibited by modern biotechnology in the past 10 years. So, biotechnology can be said to have some of the oldest as well as the some of the newest innovations that molded society's use of scientific informality.

The history of biotechnology dates back to the advent of agricultural practices over 10,000 years ago in almost every continent inhabited by humans. Evidence of selective breeding has been discovered in almost every culture that settled to build permanent communities. Middle Eastern cultures were breeding crops such as barley, flax, lentil, various peas, and vetch around 8500 B.C. Greeks adopted crop production by 7000 B.C. and agriculture then reached northwestern Europe through southeastern and central Europe by 4800 B.C. Rice production also took place around 6800 B.C. in East Asia and India. Central and South American people were cultivating corn and squash around 7000 B.C. Potatoes were bred in Peru around 3000 B.C. Cattle, fowl, goats, horses, pigs, and sheep were also bred as far back as 8,000 years ago. Many of the modern cattle breeds found today go back to beef cows and dairy cows that were selectively bred around 2000 B.C. The first record of commercial dairy farms dates to 4000 B.C.

Many historians view the birth of biotechnology with the first strategies for using microorganisms to produce foods and commercially important products. They date the origins of biotechnology to 6000 B.C. At that time, the Sumerian and Babylonian cultures used the anaerobic respiration of yeast to make beer. It is believed that civilizations in Asia and South America were using fermentation to make other types of beverages and foods at least 2,000 years before beer was brewed. In 4000 B.C. the Egyptians used the aerobic respiration of yeast to leaven or rise bread. In addition, the use of molds to flavor and preserve cheese was also being developed around 4000 B.C.

Bacteria were also being put to work in ancient times. Milk from cows and goats was preserved by using lactobacillus bacteria to prepare yogurt starting around 4000 B.C. Biotechnology did not see much growth until around 500 to 400 B.C. During this period, Mediterranean people used microbial secretions and salts to halt the metabolic pathways in meats that caused food spoilage. This led to the processes of curing and pickling that are still practiced today. During that period the idea that humans are subjective to the principles of selective breeding was recognized by the Greek philosopher Socrates who lived from about 470 to 399 B.C.

The Greek physician Hippocrates explained the mechanism of Socrates' observation around 400 B.C. stating that inheritance was partly determined by the male's contribution of something passed along in the semen. He also hypothesized that the female somehow contributed to approximately half of the offspring's traits. These observations led to a more rational strategy for the selective breeding of valuable animals and plants. Aristotle supported this hypothesis and taught it to his students around 300 B.C. Various herbal remedies were being developed at that time and medical practices required knowledge of body functions. Greek thought replaced much of the scientific ideas developed throughout Europe and the Middle East as the Greek empire grew.

The birth of the Roman empire displaced much of the Greek philosophy around 100 A.D. Technology related to building and warfare replaced biotechnology developments under Roman rule. As a result, biotechnology in Europe stagnated and remained unchanged for almost 1,000 years. This period was a part of the European Dark Ages in which scientific thought progressed slowly. The growth of biotechnology was hindered until the 1700s by the European belief in spontaneous generation. Spontaneous generation explained how organisms emerged from nonliving matter. For example, it was believed that cockroaches and rats were produced by filth and bad habits.

Much of the growth of biotechnology after A.D. 1300 involved the distillation of a variety of alcoholic beverages from fermented grain. The production of fermented products stopped in Egypt and Persia as Islam spread throughout the Middle East. This also spread to parts of Europe before the Crusades. They continued to use yeast to make breads and cereals. This practice spread throughout Africa. A greater variety of grains permitted the development of many types of breads and cakes that suited the diets of each culture that adopted leavening. By the 1500s, Europeans continued using biological processes to preserve a variety of foods leading to the development of sauerkraut. In the Americas, the Aztecs and nearby civilizations cultivated spirulina algae as a source of food and animal feed.

The invention of the microscope in the late 1600s accelerated the growth of biotechnology in Europe. This permitted Europe to outpace other nations in the growth of scientific thought. The curiousity to understand cell structure encouraged the use of analytical chemistry techniques to study cell function. In the early 1700s, amateur scientist Anton van Leeuwenhoek used a microscope to provide evidence of microscopic life. Together with researcher Robert Hooke, Leeuwenhoek heralded in the era of biogenic theory. Both of their findings helped scientists recognize that microorganisms might play a role in fermentation and cells make up the structure of complex organisms. Leeuwenhoek's work also confirmed Louis Dominicus Hamm's discovery of sperm, which paved the way to the discovery of the genetic material.

By the 1800s, biotechnology developments were being advanced by interdisciplinary scientific investigations that blended biology, chemistry, and physics. Darwin's unveiling of evolutionary theory further drove the interest in using the characteristics of living organisms to fulfill human needs. At that time, the French Scientist Louis Pasteur learned how to control the metabolic processes of organisms for preserving foods and developing medicines including vaccines. He spurred a great interest in finding cures for infectious diseases. Thus grew the fields of medical and pharmaceutical biotechnology. Pasteur was also the first to show that enzymes, and not some vital force, was responsible for metabolic processes. Vitalist thinking proposes that all life processes are animated by immaterial life spirits. One application of this was the sterilization of goods using sealed flasks heated in boiling water discovered by the Italian scientist Lazaro Spallanzani. His principles of sterilization are still used in many biotechnology applications.

The period spanning 1800 to 1900 saw a strengthening of the scientific method and an industrial revolution that brought about epic changes to agricultural and industrial technologies. Science was not accepting speculation to explain the laws of nature. It was developing a more empirical or experimental basis that encouraged scientists to understand the underlying chemistry and physics explaining living properties. Many scientists were using early biotechnology principles to develop therapeutic agents and to improve industrial processes used to make commercial chemicals. Agriculture was also benefiting by a better understanding of inheritance.

A genetics revolution came about early in the 1900s. Biologists focused many of their research efforts on understanding the principles of inheritance. This was fueled by the rediscovery of Gregor Mendel's laws of heredity that he presented to the Natural Science Society in Brunn, Austria, in 1866. Mendel suggested that undetectable bits of information are responsible for the observable traits of organisms. He also stated that the substance that produced these traits was passed from one generation to the next. By the 1940s and 1950s, the study of inheritance led to the discovery of DNA and the enzymatic pathways that drive cellular respiration and photosynthesis. The clarification of gene function led to experiments by Arthur Kornberg in which his research team at Stanford University used a strand of viral genetic material to assemble an artificial DNA sequence composed of 5,300 nucleotides.

In 1957, Francis Crick and George Gamov conducted experiments on gene expression learning and discovered the "central dogma" that explained how DNA functions to make protein. The "central dogma" that they proposed suggested that the flow of genetic information starts with DNA and goes to messenger RNA (mRNA) and then to proteins. The "central dogma" paved the way for the 1959 discovery of gene function. Francois Jacob and Jacques Monod of France recognized the mechanism of gene regulation and described the regulatory components of DNA. In a period spanning 1961 to 1966, Marshall Nirenberg used a synthesized strand of mRNA to disclose the chemistry of the genetic code. The nature of DNA culminated in 1967 with the creation of the first gene mapping technique by Mary Weiss and Howard Green. Their procedure, which is called cell hybridization, is still used in many modern biotechnology applications.

The 1970s became the era of genetic engineering. This period was ushered in by the 1972 study of Paul Berg who isolated a restriction enzyme that cuts DNA. Berg then used the restriction enzyme with another enzyme called ligase to cut and paste two DNA strands together. He created the first recombinant DNA molecule. The ramifications of Berg's findings led to the development of guidelines for DNA splicing. These guidelines were written by United States National Institutes of Health. Many scientists recommended that certain types of recombinant DNA experiments should be halted until questions of safety are addressed. These concerns in time led to the Asilomar Conference held in California in 1975. The conference was held to discuss the relevant issues of recommend DNA technology.

In 1977, the first product made by a transgenic genetically organism was created. Genentech, Inc. produced a human protein called somatostatin, or human growth hormone, in bacteria. The term transgenic describes an organism that had genes from another organism put into its genome through recombinant DNA techniques. This was the first time a synthetic recombinant gene was used to clone a protein. Genetech's work gave rise to the science of bioprocessing. Their work was then followed by the insertion of the gene for human growth hormone in bacteria by John Baxter, reported in 1979. In 1982, Eli Lilly Company was able to manufacture human insulin by placing the insulin gene inside bacteria. Other genetically modified organisms that produced therapeutic compounds were created by various companies throughout the 1980s. Although it was novel for the 1980s, the production of recombinant proteins is now a common practice in modern biotechnology.

Biotechnology in the 1970s and 1980s also assisted the birth of Louise Joy Brown in England on July 25, 1978. She was the world's first successful "test-tube" baby born using in vitro fertilization. During in vitro fertilization the egg is introduced to the sperm outside of the body. The technology that made her conception possible was recognized as an achievement for furthering the production of transgenic animals. In vitro fertilization is commonly used in modern agriculture. Frozen sperm and egg are purchased, so fertilized eggs that produce desirable animals can be raised inexpensively in surrogates.

In 1980, the U.S. Supreme Court provided a large incentive to create new life forms for biotechnology applications. They ruled that genetically altered life forms can be patented. The first patent of this nature was awarded to Exxon Oil Company for an oil-eating microorganism that was used to clean up the 1989 Exxon oil spill at Prince William Sound, Alaska. By 1981 the first transgenic animals were produced by scientists at Ohio University. They achieved this by introducing genes from other animals into laboratory mice. This is now a common biotechnology procedure used on a variety of organisms.

Biotechnology took another direction with a momentous discovery made by Kary Mullis at Cetus Corporation in Berkeley, California. In 1984, he invented a technique called polymerase chain reaction (PCR) for multiplying DNA sequences outside of a cell. The patent for PCR was sold to Hoffman-La Roche, Inc. in 1991 for $300 million. PCR remains a very important technique used in various forensic and medical biotechnology applications. PCR opened the door for the cloning of synthetic genes and fueled the production of many transgenic animals and plants. In 1986, the first genetically modified tobacco plant was grown in farm fields. The United States Environmental Protection Agency approved the release of the genetically altered tobacco plants. This was followed in 1987 by the release of genetically modified bacteria into soil and water. The bacteria were altered so that they could degrade toxic pollutants.

Another era of biotechnology was born in 1989 with the commencement of Human Genome Project. It was managed under the direction of the National Center for Human Genome Research headed by James Watson. The center was awarded $3 billion from the U.S. government to fund an effort to map and sequence all human DNA by 2005. The Human Genome Project prompted genome studies on a variety of organisms including those used in agriculture and research. A variety of new biotechnology instruments and novel laboratory procedures were fueled by the Human Genome Project. It was also responsible for birth of supercomputers used in bioinformatics. Bioinformatics is the collection, organization, and analysis of large amounts of biological data using networks of computers and databases.

The public's desire for investments in high-tech companies in the 1990s encouraged a rapid growth in biotechnology companies and products. Thousands of biotechnology products that are still being used today were developed or researched during the 1990s. Other developments in the 1990s and 2000s include the first human embryo cloning studies performed at George Washington University. Researchers were able to clone and maintain human embryos in laboratory culture dishes for several days. Their project inflamed protests from critics of genetic engineering and incited fears of unregulated human cloning. This public distaste for cloning did not stop recent endeavors to clone agricultural animals such as cattle, pigs, and sheep. The sheep named Dolly was famous for being the first mammal to be cloned from an adult cell. This feat was done by the Roslin Institute in Scotland.

The 1990s and 2000s may become the era of gene therapy. Researchers launched the field of gene therapy in 1990 when the first patient received genetically altered cells to treat a human disease that weakened the immune system. This successful attempt at gene therapy was expanded to other genetic disorders of animals and humans. Contemporary biotechnology is also making it possible for the development of pharmacogenetic treatments that are tailored to a person's genetic material. These treatments will provide more benefits and reduced risks over traditional drugs. Animals raised to produce human blood, milk, and transplant organs are already in development. In addition, plants are being developed to reduce the need for animal flesh as a source of protein. DNA information is being made simpler to decipher with the recent invention of microarrays and other procedures that identify the genes that are functioning within a cell. It is already possible to do a rapid DNA analysis on a person and a gene analysis to investigate any pending medical conditions that cannot be measured using current technologies.

BIOTECHNOLOGY INNOVATIONS

The techniques used in modern biotechnology will be briefly highlighted in this section. A basic knowledge of their fundamental principles and applications is important for understanding the way biotechnology is used to benefit life. The techniques used in modern biotechnology can be divided into three major categories: genomics, proteomics, and metabolomics. Genomics is described as the study or use of genes and their functions. It involves techniques that investigate or make use of DNA. Proteomics is the study or use of the structure and function of proteins. It includes the various ways that proteins work individually and interact with each other inside cells. Metabolomics is the study or use of specific cellular processes carried out inside and outside of a cell. It includes the interaction of the cell with other cells (physionomics) and with environmental factors (environomics).

Biotechnology techniques can also be classified into categories that analyze or apply the genomics, proteomics, and metabolomics. Analytical methods are used to analyze the function and structure of DNA, proteins, or metabolic pathways. These techniques today rely in laboratory instruments that detect the activity and chemical configuration of biological molecules. It was founded on the science of analytical chemistry, which is the analysis of chemical samples to gain an understanding of their chemical composition and structure. Application methods vary greatly and involve specific techniques for each category of biotechnology. Genomics applications require the modification of DNA. Procedures that control the functions or alter the structures of biological molecules are used in applications of proteomics and metabolomics. The manipulation of an organism's genomics, proteomics, and metabolomics is one of the fastest growing areas of biotechnology. New techniques are being developed every year for a variety of applications ranging from agriculture to pharmaceutical production.

Genomic Analysis Techniques

A procedure called karyotyping was the earliest genomic technique that provided detailed information about DNA. Karyotypes are photographs of chromosomes taken through a microscope. Cells used in karyotyping are cultured in a medium that stops them from replicating at the metaphase stage of mitosis. At this phase the chromosomes are readily visible and each chromosome can be seen separately. In addition, during metaphase, chromosomes are condensed and take on the appearance of small "X's." The chromosomes can be stained with a chemical such as trypsin-Giemsa dye that produces a predictable banding pattern on the chromosome. The pattern of bands is diagnostic for each chromosome and allows scientists to identify large abnormalities of the chromosomes. Visualization of the banding pattern using a variety of dyes is sometimes called chromosomal painting.

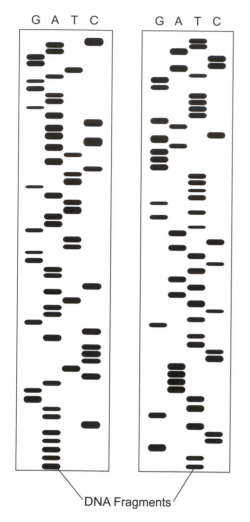

G A T C   G A T C

DNA Fragments

Figure 4.1 DNA sequencing is a genomic analysis technique used to distinguish the nucleotide sequences of DNA samples. (*Jeff Dixon*)

Details about the DNA sequence making up the genetic code was achieved with the invention of DNA sequencing. The first strategy for sequencing DNA required separating the DNA into a single-stranded piece. Fragments of the complementary strand are then synthesized using the information coding strand of the DNA. Four different chemical reactions are set up using radioactive nucleotides called dATP, dCTP, dGTP, and dTTP. An enzyme called DNA polymerase is then added to each reaction to build a complementary chain. However, growth of the chain is stopped at various points with the addition of chemicals called dideoxynucleotides. Dideoxynucleotides block further elongation to produce fragments of the complementary DNA strand. A different dideoxynucleotide is added to each tube. All the fragments made in each tube are then placed and analyzed in an electrophoresis sequencing set where the fragments are separated by size. The scientist then can calculate the sequence of the DNA by analyzing the radioactive nucleotide composition of the fragments.

Cycle sequencing is a modification of the traditional sequencing method. It also uses dideoxynucleotides to create a set of DNA fragments. However, it differs from traditional sequencing in that it uses a DNA polymerase that works even when heated to 95°C. This high temperature removes the fragments from the DNA and permits the creation of many copies of the fragments. The reactions are heated and cooled over and over again in cycles. In this method the DNA is labeled with special nucleotides that can be analyzed using spectroscopy. This is a good method for analyzing small amounts of DNA. Genomic DNA can be amplified or made into multiple copies for other types of analyses using a similar procedure called polymerase chain reaction or PCR. This process allows researchers to produce millions of copies of a particular DNA sequence within a couple of hours. PCR can be used to copy a single gene or all of the genes of an organism.

PCR uses a high-temperature DNA polymerase called Taq polymerase that can build copies of genomic DNA at high temperatures. PCR produces copies of DNA in three steps. These three steps are repeated for about 30 or 40 heating and cooling cycles. Cycling is carried out in an automated PCR thermocycler. The thermocycler rapidly heats and cools the PCR reaction mixture. The first step is a heating action that separates the DNA strands. This process is called melting. A chemical called a primer is added to the next step which is operated at a cooler temperature. Primers bind to the single-stranded DNA and act as the starting point for DNA replication with the Taq polymerase. In step three the Taq polymerase adds dideoxynucleotides to melted DNA to produce double-stranded copies of the original DNA. A procedure called "real time" PCR uses a thermocycler attached to spectrophotometer that can measure the amount of DNA copies being produced as they are made. In situ hybridization (ISH) is a technique blended with PCR in which the primers combine or hybridize in a cell or tissue. This permits the PCR reaction to be carried out within cells.

PCR is very useful in forensic science when only a small sample of DNA is found in blood, skin, or sperm samples associated with a crime. In addition, certain types of PCR are used to alter the DNA as a tool of understanding potentially harmful genetic changes called mutations. The importance of making accurate copies of the original DNA cannot be understated. Inaccurate copies can give the wrong information when trying to identify and match two or more DNA samples. A technique called nested PCR uses two pairs of PCR primers to copy a certain segment of DNA. The first primer copies the intended DNA sequence and the second primer binds within the sequence. This produces a second PCR product that is shorter than the intended DNA sequence. The second primer indicates that correct DNA sequence is being copied. There is a low probability that both primers will bind to the wrong DNA sequence during amplification.

Sometimes it is not possible to have a sample of DNA for PCR. This is true for DNA sequences that have not yet been identified. A technique called reverse transcriptase-PCR (RT-PCR) was developed to make a copy of DNA for amplification using PCR. At first, a sample of RNA that represents a particular gene is extracted from a cell that is exhibiting the characteristic of the gene. The RNA is then placed in a reaction mix in preparation for the first RT-PCR step. This first step of RT-PCR is called the first strand reaction. In this step a copy of complementary DNA or cDNA is made from the RNA using dideoxynucleotides and an enzyme called reverse transcriptase. Reverse transcriptase is capable of building a copy of DNA from an RNA template. The mixture is combined with a DNA primer in a reverse transcriptase buffer for an hour at 37°C. The cDNA is then placed into a regular PCR reaction where it is amplified. A new technique called in situ RT-PCR permits scientists to carry out the same process inside of a cell or tissue.

Restriction fragment length polymorphism (RFLP) is a method that uses proteins called restriction enzymes to chop the DNA into specific fragments based on the base pair sequence of the DNA. Specific restriction enzymes produce an RFLP sequence of DNA that has a restriction site at each end of a target sequence. A target sequence is a fragment of DNA that can be bonded to a probe made of the complementary base pairs. A probe is a sequence of single-stranded DNA labeled with radioactivity or an enzyme so that the probe can be detected. A particular DNA sequence of an RFLP is identified when a probe having a complementary sequence binds to the RFLP. The RFLP and probe mixture is placed on an agarose electrophoresis system that uses an electrical charge to separate the various RFLP segments based on their size. A technique called Southern blotting is then used to transfer the RFLP segments to a special membrane made of nitrocellulose that permits detection of the probes bonded to a particular RFLP segment. This method permits the DNA to be analyzed in chunks. It is sometimes called shotgun sequencing and is used to study DNA fragments of 2,000 base pairs to 10,000 base pairs in length.

Another strategy for sequencing large amounts of DNA is chromosomal walking. This involves the formation of two genomic libraries from the same sequence of DNA. Each genomic library is created by using a different restriction enzyme to cut the DNA into segments. Restriction enzymes called EcoR I and Sal I are commonly used for human DNA samples because they produce medium-sized fragments that are simple to analyze. A special probe is then made for binding to a particular gene. This probe is made by collecting RNA from cells that are actively showing the traits of the gene. The probe is then added to each set of genomic fragments. Fragments are then analyzed using agarose electrophoresis and Southern blotting which shows researchers the location of the gene on each set of genomic DNA samples. The samples are compared to each other and to RFLP information to determine the precise location of the gene on the DNA.

Genomic Expression Techniques

Genomic expression techniques involve strategies that detect a particular gene's activity. The activity of a gene is called expression. These are important techniques for understanding how altered genes cause illnesses and how certain factors control the expression of a gene. Certain genes called structural genes produce RNA, proteins, and other molecules as they are being expressed. Other genes called regulatory genes may only produce RNA, if any product is at all produced. The simplest method for identifying protein expression is by monitoring the levels of proteins extracted from the cells. This is done by dissolving the proteins in a solution that is analyzed using spectrophotometry. Spectrophotometry uses light to indicate the concentration of the protein. However, it is not valuable for identifying a particular type of protein.

Polyacrylamide gel electrophoresis (PAGE) is used on proteins as a way of providing information about the protein's electrical charges, shape, size, and weight. This technique uses an electrical field to separate charged biological molecules based on various characteristics of the molecules. A procedure called Western blotting is used to transfer the proteins separated by electrophoresis onto a nitrocellulose membrane. The membrane is then dipped into a solution of special probes made out of antibodies. Antibodies are proteins produced by the immune system. They are designed to attach onto specific regions of biological molecules. Antibody probes are used to identify a specific type of protein or a category of proteins having a common characteristic.

There are various other molecules that assist with detection of probes attached to the particular protein. Amino acid analyzers provide more information about the proteins associated with gene expression. These analyzers degrade the protein in amino acid units that are then detected with a modified spectrophotometer. Immunofluorescent probes glow when exposed to ultraviolet light or a chemical reaction using the molecule adenosine triphosphate (ATP). They can be detected either by using a spectrophotometer or by viewing the membrane under a microscope. In situ immunofluorescence permits the visualization of a protein as it is being produced by a cell. It can be used to indicate the amount and the location of a protein being expressed in the cell or in a tissue.

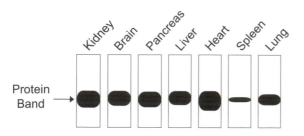

Protein Band →

Figure 4.2 Northern blotting is one genomic analysis technique commonly used in biotechnology investigations. This blot shows the amount of a particular type of RNA produced by various body organs. (*Jeff Dixon*)

A technique called Northern blotting is used to identify RNA produced during gene expression. The RNA is removed from a cell that is expressing the appropriate gene. PAGE is then used to separate all the RNA that was being produced in the cell. The RNA in the electrophoresis gel is then transferred to a nitrocellulose membrane. An RNA probe is then hybridized to the expressed RNA to identify a particular gene. Unknown sequences of RNA can be removed from the membrane and analyzed using a nucleic acid sequencer. RNA can also be characterized using ELISA antibody probes that attach to specific sequences of RNA nucleotides.

In situ hybridization can be modified to directly detect the activity of genes on an organism's genome. This is a useful method for DNA segments that do not produce proteins and make small amounts of RNA that are difficult to isolate. The technique is called fluorescent in situ hybridization or FISH. FISH uses a piece of single-stranded DNA probe that is complementary to the DNA of a gene. The probe can only bind to the gene when that particular region of the genome is being expressed. Expressed DNA is usually opened up exposing the gene to the FISH probe. A fluorescent molecule attached to the probes permits it to be visualized in the cell. This produces a fluorescent spot at the location of the gene when viewed under a fluorescent microscope. Probes can be designed to glow different colors as a way of viewing the expression of several genes at once. The technique is important for determining the way genes work together during the expression of complex characteristics.

Proteomics Techniques

In its simplest form, proteomics is the measure of protein function associated with gene expression. Many of the procedures used in genomic gene expression are also used to assess proteomics. PAGE and Western blotting are the starting points of conducting a proteomic analysis. Sophisticated types of electrophoresis, such as fluorescent 2-dimensional electrophoresis, rapidly isolates and labels a particular protein being expressed. Proteins can be extracted from the gels using various types of procedures called microdissection. Microdissection today is done using a laser-guided robot that finds the labeled protein and precisely cuts it out of the gel without accidentally capturing nearby proteins. It is then possible to carry out various other analyses on the protein.

Immunoprecipitation is a technique used to collect different types of proteins expressed in a cell at a particular moment. Cells expressing a particular trait or overall characteristic are placed in a detergent solution that forms holes in the cell membrane. The cells are then flushed with a solution containing special beads that contain antibodies for attaching to particular proteins. A centrifuge is then used to separate the beads from the cells and any molecules not attached to a bead. The proteins are then washed free of the beads and analyzed using electrophoresis. This procedure provides information about selected groups of proteins being produced during the expression of a trait.

A procedure called sequence analysis of functional domains compares the similarities of proteins between different organisms. It detects the presence of a protein component called the function domain. The functional domain is a region of the protein that permits the protein to carry out a particular job in the cell or in the body. Proteins are collected from a cell expressing a particular trait and are degraded and sequenced using an amino acid sequencer. The amino acid sequences of the functional domains are then studied to better understand the role of a protein in expressing a trait. Proteins with similar domains are then compared among different organisms to see if those proteins have similar or dissimilar roles. Certain pharmaceutical compounds are made by adjusting the function domains of particular proteins so that they have medical value.

A technique called the yeast two-hybrid (Y2H) method is used to investigate the interactions between proteins during gene expression. Although it is not a perfect method, it does allow high throughput screening of protein interactions, which is one critical component of proteomics. The Y2H method produces a colored product when two proteins interact in a yeast cell genetically altered to express the two proteins. The genes are labeled with other genes called reporter genes. A reporter gene produces some type of signal that indicates expression of the gene it is attached to. The two reporter genes used in Y2H produce a color when they interact. Expression of only one of the two genes produces no color. This technique can be expanded to investigate the interaction of complex gene groupings.

A procedure called isotope-coded affinity tagging compares the relative protein expression between two different cells or cells placed under dissimilar conditions. It is often used to distinguish between the protein expression of healthy and diseased tissue. The procedure involves a four-step process; using isotope coded affinity tags is a three step process. In the first step the amino acid cysteine in a protein is attached to a chemical called an affinity tag. Next, the tagged proteins are digested with enzymes. Tagged proteins are then separated using affinity chromatography that binds to the proteins with the tagged cysteine. A special type of affinity chromatography called biotin-avidin, or biotinolated, binding is generally used to capture the tagged proteins. The tagged peptides are the separated using ion-exchange chromatography and further characterized with spectrophometry.

In certain proteomic studies, it is important to remove a gene from the cell to study it in isolation. This involves a variety of techniques called expression system methods. Some scientists use the traditional term cloning. Cloning originally described a process in which a gene was inserted into a microorganism as a way of rapidly replicating the gene. Bacteria and yeast are the most common organisms used as expression systems. They are simple to grow and rapidly make copies of a gene inserted into their genome. Bacteria and yeast that are genetically modified with the insertion of one or more genes of another organism are called recombinant or transgenic expression systems. Bacterial expression systems are simple to produce compared to those using yeast. However, yeast are preferable expression systems for animal, human, and plant research because they have eukaryotic cells that more truly express the inserted gene.

The circular genetic material of bacteria is simple to remove, modify, and reintroduce into another bacterial cell. In addition, it is possible to successfully recombine bacterial DNA without removing it from the cell. It is also possible to insert DNA in a small circular piece of DNA called a plasmid, which is found in many bacteria. Plasmids are not always replicated with the bacterium. Certain procedures require the placement of a large piece of DNA called a bacterial artificial chromosome (BAC) in bacteria. However, they are naturally transmitted from one bacterium to another by a process called conjugation. Yeast cells are usually genetically modified by adding an artificial plasmid or by inserting a small synthesized piece of DNA called a yeast artificial chromosome (YAC). Any gene inserted into a microorganism must be placed into a sequence of DNA called an expression vector. The original term for expression vector was cloning vector.

Scientists have created many types of expression vectors for a variety of organisms and for a wide range of purposes. A typical expression vector is composed of two insertion points, a promoter, a regulatory region, the gene of interest, and a terminator region. Insertion points, or sticky ends, are single-stranded sequences cut with restriction enzymes. These ends are designed to lock the expression vector into a particular region of a DNA in an artificial chromosome, ge-

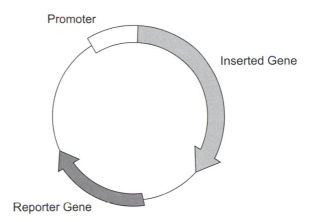

Promoter

Inserted Gene

Reporter Gene

Figure 4.3 New genes are inserted into cells by first placing them on sequences of DNA called expression vectors. They facilitate the expression of the new gene in the cell recipient cell. (*Jeff Dixon*)

nome, or plasmid. It is important that the expression vector in a region of DNA that does not disrupt cell function by damaging or interfering with other genes. A promoter is a region of DNA that identifies the location of a gene. It is the site where gene expression begins. Scientists have developed many types of promoters specific to requirements of gene expression for different types of cells. Many promoters are derived from the DNA of viruses.

The regulatory region, or operator region, of the expression vector is the section of DNA that permits control of the gene expression. Certain regulatory regions are constitutive meaning that they automatically express the gene once it is inserted into the vector. Inducible regulatory regions express the gene in the presence of some type of cell signal or environmental factor. It is common for the researcher to make a selective inducible regulatory region that is expressed only when a certain chemical is added to the cells. Repressible regulatory regions prevent gene expression under certain conditions. They are commonly used when investigating gene interactions that involve the fine control of trait expression. Reporter genes are regularly added to the regulatory region as a way of ensuring the correct functioning of the expression vector. The green fluorescent protein gene from jellyfish is a common reporter gene. Its protein product glows green when exposed to blue and ultraviolet light.

The gene or genes being inserted in the expression vector can be placed forward or backward. A forward gene expresses the trait.

However, sometimes it is desirable to produce a backward expression for studies investigating the regulation of traits. This reversed gene produces an antisense RNA molecule. Eukaryotic genes are usually not expressed correctly in prokaryotic cells. The DNA of eukaryotes contains regions called intron information that must be edited from the RNA by molecules called small nuclear ribonucleoproteins (snRNP). Prokaryotic cells cannot edit the introns because they do not produce snRNPs. Therefore, eukaryotic DNA placed into bacteria must be modified to remove introns. This is done by using reserve transcriptase to make a complementary DNA from RNA that was edited by the appropriate snRNP.

A terminator region is needed to discontinue each round of expression. It signals the end of the DNA comprising the trait information. Various terminator regions are available for the different cells for proteomic work. Protein labeling is another important factor for gene expression. Sometimes it is essential to modify the gene so that it has a small sequence that produces an amino acid label. This label signals the cell on how to process the enzyme. Some gene products must be modified by the cell before it can be used in trait expression. Other labels are needed to determine where the gene product is being transported in the cell. Some expression products must remain in the cell, while others have to be secreted to the outside of the cell.

Metabolomics

Metabolomics by far involves the widest array of techniques needed to investigate various aspects of gene expression. It has to measure the interaction of various genes that contribute to the characteristics of an organism. A typical metabolomic study requires the use of several analytical instruments that detect the host of biological molecules involved in even the simplest metabolic pathways. Laboratory information management systems are often needed to blend and interpret the data coming in from a variety of analytical instruments monitoring a cell's or a tissue's metabolic processes. Metabolomic studies can also provide information about physiomics and environomics. This is achieved by placing the cells under growing conditions that favor the expression of a particular metabolic pathway.

Fluorescence Activated Cell Sorting (FACS) is one way to distinguish between two metabolically different cells. It uses a laser light to identify differences in cell—the appearance or phenotype of cells having dissimilar patterns of gene expression. A mixture of cells are passed through the laser beam and are then sorted based on a particular pattern of proteins that make up the cell's structure or metabolic activities. This technique is commonly used to identify if cells are carrying out similar tasks under different environmental conditions. More accurate results can be achieved with protein arrays. Protein arrays us protein-based chips that bind to proteins extracted from a cell. The array is composed of a small surface covered with antibodies attached to fluorescent tags. Cell proteins that bind to the array take on the tag which is then measured using a laser light that scans the surface. It provides fluorescent patterns or fingerprints that reflect the metabolic activities of a cell. ELISA testing can also be done when looking for a small number of proteins involved in simple metabolic pathways.

DNA microarrays are increasingly becoming important metabolomic tools. This technology permits scientists to investigate the orchestrated activity of thousands of genes involved in metabolic functions. Traditional genomic and proteomic methods work on a "one or two genes in one experiment" basis. This means that the information must be gathered very slowly. In addition, these traditional technologies make it difficult to study all of the complex gene interactions taking place in a cell or tissue. Microarrays provide a whole picture appropriate to the gene regulation taking place during a particular metabolic reaction. They do this by reacting with thousands of RNA molecules at the same time and producing a genetic fingerprint of a particular cell activity. The activity of a particular gene is indicated by a color change on the microarray device.

RNA interference (RNAi) involves a variety of methods used to block the expression of a particular gene product. It permits researchers to obstruct the function of a gene involved in a metabolic pathway. This is useful information for carrying out metabolic engineering and for studying metabolic diseases. In its simplest form, double-stranded RNA that is complementary to the sequence of a targeted gene is manufactured in the cell. This is accomplished by using one of several ways of adding an antisense RNA for the transcribed RNA. Antisense RNA then binds to the expressed RNA, preventing further expression of the trait by blocking protein synthesis. Small segments of antisense RNA, called small interference RNA (siRNA) sequences, can be added to a cell to produce partial RNA hybrids that are destroyed by the cell using an enzyme called dicer. Dicer chops, or dices, the RNA hybrid. Cells use dicer to protect against RNA introduced by viruses attacking the cell. Other forms of RNAi are being developed that prevent RNA from doing its job in a variety of ways.

Another metabolomic strategy involves altering the DNA in a variety of ways to produce a genetically modified organism (GMO). A common practice in metabolomics involves the production of a knock out. A knock out is a cell that is genetically modified so that one or more particular genes are not permitted to function. Eukaryotic cells have two copies of each trait. Therefore it is essential to have both copies of a particular gene deactivated or silenced. Knock outs can be produced by removing the gene or inducing a mutation that disables its expression. Many of these mutations are induced on a regulatory region of that particular gene. Some regulatory regions can directly control several genes. Therefore, caution is used when trying to disable a gene by altering regulatory DNA segments. Another strategy for knocking out a gene involves replacing the coding portion of the gene with an inactive allele.

A method called homologous recombination involves the replacement of a gene with an engineered DNA called a construct. This method ensures that the inserted gene fits exactly in the spot with the gene being removed and replaced. Other methods of genetic recombination alter the sequence of genome and may interfere with other genes. Homologous recombination requires knowledge of the DNA sequence of the gene being replaced. Restriction enzymes are selected to excise the original DNA without cutting into nearby genes. With this information, it is possible to replace any gene with a DNA construct of your choice. The method has a few more details than will be illustrated here, but the essential information is retained.

PRODUCTION OF GENETICALLY MODIFIED ORGANISMS

Genetically modified organisms (GMOs) are produced for a variety of agricultural, commercial, and medical purposes. Certain GMOs are solely made for bioprocessing while others are created to study human disorders. Many GMOs are cultured in bioreactors while others are grown in farm fields. The earliest GMOs were genetically modified bacteria produced for research curiosity in 1973. They were created by inserting a synthetic gene sequence into a plasmid that was then introduced into a bacterium grown in culture. The new plasmid was made by cutting open its DNA with restriction enzymes and inserting a novel DNA sequence using an enzyme called ligase that combines nucleotides. These plasmids were inserted into the bacteria by using a natural process called transformation in which bacteria can take up whole pieces of DNA without digesting it into nucleic acids. Certain conditions must be maintained for transformation to take place. In addition, much of the DNA taken is degraded in the cytoplasm before it can be expressed. The chloroplasts and mitochondria of eukaryotic cells can be genetically modified using many of the techniques used for bacteria.

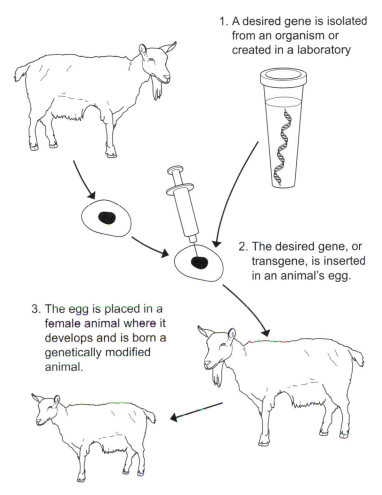

1. A desired gene is isolated from an organism or created in a laboratory

2. The desired gene, or transgene, is inserted in an animal's egg.

3. The egg is placed in a female animal where it develops and is born a genetically modified animal.

Figure 4.4 Transgenic animals are produced by inserting the genes from an unrelated organism into an egg from the animal. The resultant offspring then has unique genetic characteristics from the original animal. (*Jeff Dixon*)

More effective insertion methods were developed over time to ensure a better success rate when producing GMO bacteria. Scientists began to exploit the fact that certain viruses called bacteriophages insert DNA into the cytoplasm of bacteria. This process is called transduction. As a result, artificial sequences of DNA were placed into bacteriophages that then served as insertion vectors. Certain types of insertion vectors were developed that encouraged the insertion of the DNA sequence into the bacterial genome. This involved the use of lysogenic viruses that naturally place their DNA into the genome of bacteria. Another method called conjugation was exploited to produce new GMOs. During conjugation, the bacteria replicate and insert a plasmid into another bacterium. They transfer the plasmid by producing a temporary tube connecting the two bacteria.

It is more difficult to genetically modify eukaryotic cells because their genomic DNA is hidden away in the nucleus where gene expression must occur. Plasmids would normally be destroyed in the cytoplasm by enzymes that protect against viral invasions. In addition, most eukaryotic cells carry two sets of each trait. Hence, it is important to disable both sets of genes that could interfere with expression of the inserted genes. Yeast were the first eukaryotic GMOs because they are single celled and have only one copy of each gene. Recombinant DNA is introduced into yeast using a variety of techniques that are now commonly used on many eukaryotic cells. Animal and plant cells are now commonly genetically modified for agricultural uses, bioprocessing, and even gene therapy in humans. Most eukaryotic cells do not naturally carry out transformation. However, they can be induced to take up DNA from the environment through a process called transfection.

Transfection can be achieved using heat shock. Heat shock involves the incubation of cells and an expression vector in a solution containing calcium ions at $0°C$. The temperature of the medium is then rapidly elevated to $40°C$. This produces a heat shock effect that causes some of the cells to take up the vector. Heat shock works very well for animal cell transfection. Electroporation is a transfection method in which cells are exposed to a high-voltage pulse of electricity. This causes the cell membrane to form temporary pores that allow an expression vector to enter the cell. This process is useful for a variety of cells. Another transfection method called chemoporation forces openings in the cell membrane using treatments with soap-like solutions.

Viruses and certain types of bacteria can also be used for transfection. The vector is first incorporated into the virus or the bacterium that in turn inserts the vector into the cells. DNA can be inserted using plasmids or vectors that insert into the genomic DNA. Viruses used for transfection must be genetically modified so that they cannot replicate or harm the cells. Adenoviruses and retroviruses are the most commonly used viruses for transfection. The adenovirus, or cold virus, inserts the vector into the nucleus where the genes are automatically expressed. However, the genes are not inserted into the DNA and therefore are not replicated as the cell divides. This process was used in the first human gene therapy trial where a normal gene was inserted into blood cells to cover up the effects of a defective gene. Retroviruses for transfection contain an RNA copy of the gene. The retrovirus then makes a DNA copy of the gene when it infects the cell. This copy can then be incorporated in the cell's genome meaning that it will be passed along during cell division. Viruses are useful for delivering DNA to many cells in the body of any organism. They can be designed to enter only select cells such as those of the liver or the lungs.

A common transfection method for plant cells uses a bacterium Agrobacterium tumefaciens. Agrobacterium tumefaciens normally causes tumor-like diseases in plants. Cells from a plant are removed and cultured in a nutrient medium. These cells then undergo division and produce a mass of plant tissue called a callus. Callus is a mass of stem cells that has no particular role. However, these stem cells can be induced to produce roots and stems when subjected to a particular recipe of plant hormones and nutrients. Most GMO plants are grown from callus cultures that were transfected using genetically modified Agrobacterium tumefaciens. *Agrobacterium tumefaciens* carries a piece of DNA called the Ti plasmid. The Ti plasmid can be removed and modified into an expression vector. It is then placed back into the Agrobacterium tumefaciens which inserts the Ti plasmid into the callus cells that it infects. The callus cells are then grown into a plant that hopefully expresses the new genes.

Other methods for producing eukaryotic GMOs include bioballistics, microinjection, hybridoma formation, and liposome fusion. Bioballistics uses a machine called a gene gun to introduce DNA fragments or expression vectors into a cell. The gene gun uses compressed air to shoots microscopic gel, gold, or plastic particles coated with DNA at the cells. It is very effective for fungi and plants whose cells are covered with a cell wall. The cell wall sometimes reduces the effectiveness of transfection methods because it can act as a barrier that blocks DNA uptake. Gene guns can produce a powerful enough force that can deliver the DNA right into the nucleus. Bioballistics is useful for transfecting many cells at a time.

Microinjection uses a microscopic needle or pipette to inject DNA directly into the nucleus. This method is commonly used on embryos and stem cells. It has been used to produce transgenic animal zygotes that were then successfully implanted into the uterus of a surrogate mother. These fetuses then developed into normal animals with the novel DNA incorporated into the chromosomes of every cell. Microinjected zygotes cells can also be cloned using a technique called embryo splitting.

Embryo splitting involves the separation of cells from a developing animal or plant. These cells have stem cell capability and can grow up to be identical individuals. In effect, it is an artificial way to produce twins. It is possible to produce up to sixteen copies from one microinjected zygote. Embryo splitting was first done on frogs in the 1950s and on cattle in the 1970s.

Hybrid cell formation uses chemoporation or electroporation to fuse two genetically different cells into one completely combined cell. The first hybrid cells were hybridomas. Hybridomas were made by collecting an antibody-producing cell from an animal. These were then fused with a cancerous immune cell called a myeloma. The hybridoma cells are then cloned and tested for the production of a desired antibody. Hybridomas readily clone themselves and can survive for long periods in a culture and in the body of an animal. Many types of hybrid cells have been made by combining a variety of different organisms. However, these cells are usually not capable of cloning themselves unless they have a similar arrangement of chromosomes.

Expression vectors can be enclosed in microscopic globules of fats called liposomes. Liposomes are composed of the same types of fats found in cell membranes. They are made by mixing a solution of phospholipids under special conditions that produce uniform spheres. The spheres can be mixed in a special medium containing DNA or other material that then becomes encapsulated in the liposome. Proteins can also be incorporated into the liposome membrane. These proteins are used to help the liposome carry out various tasks used in biotechnology. They can also be used to help the liposome act like an artificial cell. Liposomes can be designed to fuse with the cell membrane and nucleus of cells as a method of delivering DNA into a cell. This works for many types of cells and is applicable for delivering genes to cells throughout the body of an organism. Liposomes are also used to transport medicines and toxic drugs into diseased cells. Researchers have recently converted red blood cells into liposomes by adding drugs and DNA to the cells.

Cloning

Three major types of cloning can be used today for the production of genetically modified organisms: reproductive, somatic cell, and stem cell cloning. Reproductive cloning is primarily used to produce genetically modified animals capable of passing along the new characteristics to its offspring. Plants, unlike animals, can easily be cloned from stem cells found throughout the plant and in the callus. Reproductive cloning

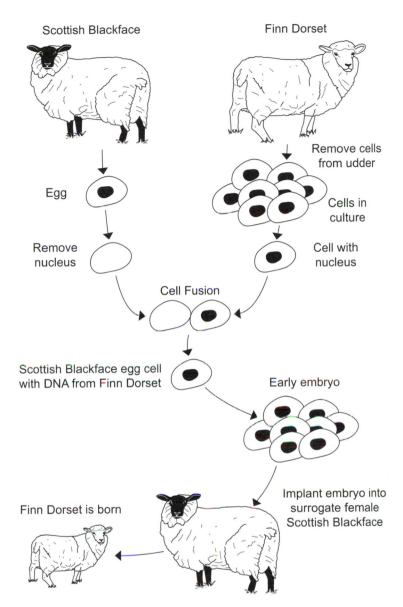

Scottish Blackface

Finn Dorset

Remove cells from udder

Egg

Cells in culture

Remove nucleus

Cell with nucleus

Cell Fusion

Scottish Blackface egg cell with DNA from Finn Dorset

Early embryo

Finn Dorset is born

Implant embryo into surrogate female Scottish Blackface

Figure 4.5 A lamb called "Dolly" was created by cloning a Finn Dorset sheep using the egg and surrogate of a Scottish Blackface sheep. Cloning is becoming a method for making copies of animals and plants with desired characteristics. Plus, it may be possible to use cloning to grow back damaged body organs. (*Jeff Dixon*)

starts with the collection of eggs from a female animal that was injected with fertility drug. Fertility drugs cause the animal to produce numerous eggs during one ovulation period. This process is commonly used in human fertility clinics that carry out in vitro fertilization. In vitro fertilization is defined as the artificial fertilization of an egg with sperm outside of the body.

The eggs are removed and are then genetically modified in one of three ways. The nucleus can be removed from the egg and replaced by the nucleus of an adult cell from the same organism. Eggs treated this way usually express the DNA of the inserted nucleus and encourage embryological development into an identical copy of the individual. The copy grows at the same rate of the original animal and matures into a similar individual. This type of cloning has been successfully done on domesticated animals such as cattle, cats, dogs, goats, mice, pigs, rabbits, sheep, and an ox called an Asian gaur. A variety of endangered animals kept in zoos are being cloned to preserve their species. Similar work on humans and monkeys has failed.

Another way of genetically modifying the egg involves replacing the original nucleus with that of a cell from the different organisms. These organisms are called chimeras because they are composed of two different organisms. The chimera is a mythical organism made up of the parts of several animals. These clones are currently produced for genetics research. A similar process is being done as a strategy to clone already extinct animals. DNA recovered from dead animal specimens can used to replace the genome in the nucleus of a related animal. The New Zealand government is initiating a project in which this type of cloning will be used to revive the extinct Huia bird.

A third form of reproductive cloning requires that the egg's nucleus is replaced with a genetically modified nucleus. The genetic modifications can be as simple as the addition or removal of one or two genes from a similar animal's nucleus. It can be as complex as the use of an artificial chromosome containing unique characteristics. This type of reproductive cloning can be used to produce chimeras that have agricultural or commercial value.

Some of the characteristics of a cloned organism change due to an imprinting effect induced by the egg's cytoplasm. Imprinting is a process by which the metabolism of the cytoplasm and the endosymbiont organelles affect gene expression. In effect, the cytoplasm of a sheep egg will cause uncharacteristic expression to a goat nucleus. Unfortunately, care must be used when selecting a donor nucleus. It has been found that the DNA of cells from mature organisms is damaged in several ways. It is common to find many mutations that can produce a variety of undesirable traits. In addition, the ends of the chromosomes, called telomeres, are degraded as cells undergo cell division through the life of a multicellular organism.

After genetic modification, the eggs are placed in a culture flask with a medium that encourages the egg to undergo embryological development. Cells of the embryo can be removed and genetically tested without harming the embryo. At this stage it is also possible to blend the cells of two different embryos to form a chimera. This process has already been done on a goat and sheep embryo to produce an organism called a "geep." An organism such as a geep is fertile. How it will pass along the sperm or eggs for either a goat or a sheep depends on whether its reproductive organs were formed from the goat embryonic cells or from the sheep embryonic cells. After a certain stage of the development, the embryo is placed into a female that has been induced into pregnancy with hormone treatments. Embryo implantation is a common practice used for in vitro fertilization in fertility clinics.

Somatic cell cloning involves the cloning of body cells, or somatic cells, to regenerate a whole organism. This process is most effective in fungi and plants because most of their cells can undergo changes that permit them to regenerate the whole organism. However, it is a difficult task to carry out this process in animals. Most animal cells are difficult to regenerate because they lose the ability of totipotency. Totipotency is the capability of a cell to replicate and then develop, or differentiate, into other body cells. In order to achieve totipotency, the cell must be able to undo its differentiated state, or dedifferentiate, and take on the genetic conditions of a stem cell. This involves complex metabolic signals that are difficult to mimic in the laboratory. Cells from simple animals such as worms will readily dedifferentiate. However, the highly specialized cells of higher animals such as insects and mammals are virtually unable to dedifferentiate.

Stem cell cloning is a simple process to carry out because stem cells have the natural ability to form tissues, organs, and whole organisms. This process is not usually used for cloning whole organisms. It is difficult to find totipotent stem cells in many organisms. Totipotential stems cells are common in the animal fetuses, but are lacking in the adults. Cells called multipotential and pluripotential stem cells are capable of regenerating tissues and organs. They are available in adult tissues and organs such as the bone marrow and skin. These cells can be harvested and cloned into replacements for damaged tissues and organs. A type of biotechnology called tissues engineering uses stem cells and synthetic materials to build artificial organs. Physicians are hoping to use stem cell cloning to replace diseased cells of people with ailments such as Alzheimer's disease, blood cell cancers, nervous system damage, and Parkinson's disease.

5

PRINCIPAL PEOPLE OF BIOTECHNOLOGY

INTRODUCTION

No one person was responsible for the birth of biotechnology. Many unknown people thousands of years ago created the agricultural and commercial practices that provided the direction for modern biotechnology developments. The principal people of modern biotechnology are from a variety of scientific disciplines. Many of the contributors to biotechnology were biologists. However, it also took the efforts of chemists, computer information scientists, engineers, medical doctors, mathematicians, and physicists to produce biotechnology innovations.

Contributions to biotechnology's development vary from the invention of specific laboratory techniques to the formulation of scientific ideas that changed the way scientists viewed nature. Many of the scientific discoveries that built modern biotechnology are usually associated with scientists working in university laboratories. Early biotechnology was predominantly performed by scientists at universities. After the 1980s it became more common for scientists working in private corporations to come up with biotechnology innovations. Equally important are the contributions of scientists who work for government agencies such as the U.S. Department of Agriculture or the Kenya Agricultural Research Institute (KARI) in Africa.

Biotechnology innovations come from many nations. Discoveries are not restricted to the wealthiest nations. Many new techniques have come out of India, Korea, and Mexico. Women have been making contributions to modern biotechnology for many years. Many important principles of DNA function and structure were investigated by women. The same is true for contributions by people of color. Advances in biomedicine that contribute to cloning and drug design were achieved by Black and Hispanic scientists. Science represents the endeavors of people coming from a variety of cultures and religious beliefs. Many of the early principles of science were developed by Arabic peoples. Scientific contributions are made by Buddhist, Christian, Islamic, and Jewish people. Unfortunately, not everybody was given equal access to science careers early in the history of modern biotechnology. As a result, most of the discoverers mentioned in this section are male Americans and Northern Europeans.

CONTRIBUTORS TO BIOTECHNOLOGY

Thousands of people throughout history have made scientific and technological discoveries that advanced biotechnology. Some people made large-scale contributions that changed the way science and technology were practiced. Many biotechnology applications came from these discoveries or inventions. Other developments were very specific and progressed on area of biotechnology. The scientific contributors described below represent the breadth of people who were somehow involved in the growth of biotechnology. Those who are included in this listing represent the diversity of people who practiced science.

Al-Kindi

Abu Yousuf Yaqub Ibn Ishaq al-Kindi was born in AD 801 in Kufah, Iraq. He came from a professional family who encouraged education and fostered inquisitive thinking. Modern biotechnology would not be where it is today without freethinking people such as al-Kindi who promoted the importance of scientific inquiry. Many of the early scientific principles adopted during the rebirth of European science in the Renaissance period were fashioned by al-Kindi's works. Al-Kindi developed a deep knowledge of Greek science and philosophy. He applied the most accurate components of Greek science to geography, mathematics, medicine, pharmacy, and physics. Al-Kindi opposed controversial practices such as alchemy and certain types of herbal healing practices that he discovered were based on weak premises. He stressed the philosophy of "empiricism." Empiricism is based on the principle that the only source of true knowledge is through experiment and observation. Al-Kindi's passion for empiricism was introduced in Europe during the era of the crusades. His philosophy gradually replaced many of the supernatural practices that dominated agriculture and medicine during the Dark Ages of Europe. Many of the great European Renaissance philosophers and scientists who heralded modern science relied on the works of al-Kindi. Some of his scientific writings were cited even into the early 1900s. Al-Kindi was persecuted for his empiricism beliefs during an orthodox uprising in Iraq from AD 841–861. Many of his writings were confiscated and destroyed during that period. Al-Kindi died in AD 873.

W. French Anderson

Dr. Anderson was born in Tulsa, Oklahoma, in 1936. He showed an aptitude for science and completed his undergraduate studies in biochemistry at Harvard College. Anderson then did graduate work at Cambridge University in England. He returned to the United States to complete a medical degree at Harvard Medical School. Anderson focused his interests on medical research and was offered a position at the National Heart, Lung, and Blood Institute at the National Institutes of Health in Bethesda, Maryland, near Washington, DC. At the National Institutes of Health, he worked as a gene therapy researcher for 27 years. Anderson is most noted for being the "Father of Gene Therapy." He investigated using viruses as a tool for transferring normal genes into genetically defective animal cells. In 1990, Anderson left the National Institutes of Health to direct the Gene Therapy Laboratories at the University of Southern California School of Medicine. The success of his research there prompted him in 1990 to form a collaborative human gene therapy trial with Michael Blaese and Kenneth Culver who were at the National Institutes of Health. Anderson and his team performed the first approved gene therapy test on a 4-year-old girl with an immune system disorder. They inserted normal genes into her defective blood cells as a treatment for the disease. The first gene therapy experiment to treat a blood disease called thalassemia was performed in 1980 by Martin Cline of the University of California at Los Angeles. However, he was reprimanded for the experiment because he did not have an approval to conduct the experiment from the college and from the National Institutes of Health.

Werner Arber

Born in Switzerland in 1929, Arber studied biophysics at the University of Geneva where he received his PhD. Early in his college education he worked in research laboratories studying the structure of biological molecules. In 1958, Dr. Arber moved to the University of Southern California in Los Angeles where he was introduced to genetics research. His research there focused on the effects of radiation on bacterial DNA. Dr. Arber then returned to Switzerland where he held professor positions first at the University of Geneva and then at the

California Institute of Technology in Pasadena. His research on the bacteria that resisted the damaging effects of DNA led to the discovery of restriction enzymes. Restriction enzymes are powerful chemical tools of biotechnology. These enzymes permit scientists to carry out modern genetic analysis and genetic engineering techniques. Without this discovery, the field of biotechnology would not exist. The significance of his findings was recognized early by the scientific community. For his diligent work, Arber was awarded the Nobel Prize in Medicine in 1978. Currently, Arber is a professor of molecular microbiology at the University of Basel. His current research investigates horizontal gene transfer and the molecular mechanisms of microbial evolution.

Oswald T. Avery

Oswald Avery was born in Halifax, Nova Scotia, in 1877. Avery had a strong religious upbringing and played cornet music at his father's Baptist church in New York City. His family had a modest income and lived in one of the poorer sections of the Lower East Side in New York City. Music was his main interest through his early college studies. Avery won a scholarship to the National Conservatory of Music. In 1893, he pursued his interest in music at Colgate University in New York. A change in interest caused Avery to study medicine at Columbia University Medical School in New York City. While there he took part in medical research and decided to make a career doing studies on bacterial diseases. Avery found research to be more intellectually stimulating for him than practicing medicine. His research on tuberculosis led to a position at the prestigious Rockefeller Institute Hospital where he studied the bacteria that cause pneumonia. In the early 1940s, Avery and Maclyn McCarty were the first to recognize that DNA transfer was responsible for the transmission of traits in bacteria. Their findings started the drive to understand the chemistry of inheritance. The research also provided a method of carrying out early attempts at genetic engineering. Avery received many international honorary degrees and awards for his contributions to genetics. He died in Nashville in 1955.

David Baltimore

David Baltimore was born in 1938 in New York City. While in high school, Baltimore took part in a summer internship at Jackson Memorial Laboratory in Bar Harbor, Maine. His experiences at the laboratory motivated him to biology. He went to Swarthmore College to study biology, did his initial graduate studies in biophysics at the Massachusetts Institute of Technology, and then received a PhD in virology from Rockefeller University in 1964. His first job was at the Salk Institute in La Jolla, California, where he performed research on viruses. Baltimore then took a professor's position at the Massachusetts Institute of Technology. He continued working on a group of viruses called retroviruses. He discovered that retroviruses contain a previously unknown enzyme called reverse transcriptase that enables them to convert RNA information into a strand DNA. This controversial discovery was contrary to current beliefs that only DNA can be used as template to build another copy of DNA. Baltimore shared the 1975 Nobel Prize in Physiology or Medicine with Renato Dulbecco and Howard Temin for their work on retroviruses. He was awarded the Nobel prize at the age of 37. Reverse transcriptase is a valuable tool in many biotechnology applications. Baltimore made many important contributions to the study of viral structure and reproduction. He made significant contributions to national policy concerning the AIDS epidemic and recombinant DNA research. Baltimore was selected to be president of the California Institute of Technology in 1997 and remained in that position through 2006.

George W. Beadle

George W. Beadle was born to a farm family in Wahoo, Nebraska, in 1903. Beadle said that he would have become a farmer if it were not for the influence of a teacher who encouraged Beadle to study science. As a student at the University of Nebraska, Beadle worked in a lab that introduced him to the study of wheat genetics. Beadle then went to Cornell University in New York to complete a PhD in genetics. He studied genetics long before much was known about the chemistry of inheritance. His college studies included working with internationally famous geneticists in America and Europe. The quality of his research earned Beadle a fellowship to do genetic studies at the California Institute of Technology where he studied fruit fly inheritance. He worked there until becoming Chancellor of the University of Chicago. In 1958, Beadle shared a Nobel Prize in Physiology with Joshua Lederberg and E.L. Tatum. The award recognized their fundamental research on bread-mold genetics. Their bread mold studies showed that genes were the unit of DNA that programmed for the production of proteins. This provided the foundation for understanding the chemistry of an organism's traits. Beadle's scientific contributions are the basis of almost every biotechnology application. He died in 1989.

William James Beal

William James Beal was born in Adrian, Michigan, in 1833. He graduated from the University of Michigan in 1859 with research interests in plant breeding. Beal had various teaching positions until he took a professorship at the State Agricultural College of Michigan in 1870. Beal had a broad area of research interests that included agriculture, botany, forestry, and horticulture. A strong proponent of Charles Darwin, Beal used the principles of natural selection to breed hardier varieties of plants. His initial breeding experiments produced a 21–51 percent increase in corn yields. Beal was the first person to publish field experiments demonstrating a phenomenon called hybrid vigor in corn. Hybrid vigor is the increased growth produced by breeding two dissimilar parents. His research built the foundation for crop testing methods used in modern agricultural biotechnology. Beal had the honor of serving as the first president for various scientific societies including the First President of the Michigan Academy of Sciences, the Botanical Club of the American Association for the Advancement of Science, and the Society for the Promotion of Agricultural Science. He was honored by having a park in East Lancing, Michigan, dedicated in his name. Beal Botanical Gardens is the oldest continuously operated botanical garden in the United States. He died in Michigan in 1924.

Paul Berg

Paul Berg was born to a Jewish family in Brooklyn, New York, in 1929. He knew he wanted to be a scientist by the time he entered junior high school. Berg wrote that he was inspired to study medicine after reading the book *Arrowsmith* by Sinclair Lewis. This interest was fostered by a high school teacher who held afterschool science activities and sponsored a science club. Berg did his undergraduate studies at Pennsylvania State University and then completed a PhD at Western Reserve University in 1952. He studied the chemistry of certain metabolic pathways while at Western Reserve University. Berg then worked at several institutions before going to Stanford University where he spent most of his scientific career. His research at Stanford University in California led to a Nobel Prize in Chemistry in 1980. Berg worked with Walter Gilbert and Frederick Sanger on the chemistry of genetically engineered proteins. Their research provided the information needed for scientists to successfully put animal and plant genes into bacteria. This technique is commonly used to produce a variety of medicines. Berg was one of the scientists who organized of the Asilomar conference on recombinant

DNA in 1975. This conference brought out many of the scientific and ethical concerns of genetic engineering. Berg understood that his research opened the door to many types of genetic engineering research. He was concerned whether all research of this type was performed ethically and safely. Berg has received numerous awards and is currently director of the Beckman Center for Molecular and Genetic Medicine at Stanford University.

Herbert Boyer

Herbert Boyer was born in Pittsburgh, Pennsylvania, in 1936. Most of the families in his neighborhood worked in mining and railroad jobs. As a youth Boyer wanted to be a professional football player. With a new career path in mind, Boyer entered college as a premed major. However, he abandoned those goals to pursue graduate work in biochemistry at the University of Pittsburgh. At first Boyer was not interested in doing research. He enjoyed doing the technical duties around the laboratory. However, he was encouraged to expand his interests and then went to Yale University to study enzyme function. In 1966, Boyer was offered at professorship at the University of California at San Francisco to do research on bacterial genetics. He was fortunate to form a collaboration with Stanley N. Cohen who was interested in altering the genetic material of bacteria. Boyer and Cohen developed a strategy for manipulating DNA that became the basis of modern genetic engineering. The commercial potential of Boyer's research spurred him to start a biotechnology company called Genentech, Inc. His company was unique for the middle 1970s because it employed genetic engineering to produce pharmaceutical products. Boyer continues to serve at Genentech on the board of directors. He was awarded numerous honors for his industry and research achievements.

Sydney Brenner

Sydney Brenner was born of British nationality in South Africa in 1927. His early college education in the sciences was done in South Africa. Brenner then did his doctoral studies in physical chemistry at Oxford University in England. It was at Oxford that he started studying the structure and function of genes working with many of the discoverers of DNA stucture. He held positons at the Medical Research Council Molecular Genetics Unit in Cambridge, England, before moving to the Molecular Sciences Institute in Califonia. Brenner is most noted for his early research that produced an understanding of protein synthesis and helped unlock the genetic code. In the 1960s, Brenner began using a roundworm called *Caenorhabditis elegans* as an experimental system for analyzing complicated gene interactions. His major interest was the genetics of neural development. During an interview he mentioned that "I'm called 'the father of the worm,' which I don't think is a very nice title." Brenner received many international honorary degrees and was awarded much recognition for most of his research. However, his earlier contributions to genetics led to a Nobel Prize in Physiology or Medicine in 2002. He shared the Noble Prize with Robert Horvitz of Massachusetts Institute of Technology and John Sulston of the Wellcome Trust Sanger Institute in Cambridgeshire, England. Brenner remains active with the Human Genome Project and continues to work at the The Salk Institute in La Jolla, California.

Pat Brown

Patrick O. Brown was born in 1954 in Washington, DC. His curiosity of science compelled him to study chemistry at the University of Chicago. He then stayed at the University of Chicago to complete a PhD in biochemistry and a medical degree. Brown stayed in Chicago to do his medical residency studies. An interest in research led Brown to investigate biochemistry and genetics as a professor at the University of California in San Francisco. In 1988, Brown joined the Departments of Pediatrics and Biochemistry at Stanford University School of Medicine. Brown's research at Stanford focused on the rapid identification of human DNA. His interest in DNA was nutured by Brown's enthusiasm for learning about the biochemistry of gene function. He was interested in expediting the pace of the newly forming Human Genome Project. In 1992, Brown developed a way of simultaneously analyzing the characteristics of thousands of minute fragments of DNA. He was eventually able to identify 40,000 DNA fragments at a time. The technology for performing this feat was called DNA microarray. A microarray is a wafer similar to a computer chip that can be used to rapidly determine the presence of particular DNA sequences. Microrray technology revolutionized biotechnology. Many related types of technologies have been developed based on Brown's original microarray. Brown has received international awards for his research achievements. His current research focuses on the identification and function of disease-causing genes.

George Washington Carver

Born a slave in 1864 in Diamond Grove, Missouri, Carver and his mother were kidnapped by slave raiders when he was an infant. Carver eventually bought his freedom and worked as a farm hand. He saved enough money for college and was admitted as the first Black student to attended Simpson College in Indianola, Iowa. Carver then earned a M.S. degree in 1896 at the Iowa State College of Agriculture and Mechanic Arts (Iowa State University). His detailed observations about crop characteristics changed the way agriculture viewed the use of crop plants. Using his knowledge of chemistry he was able to derive 300 products from peanuts and 100 products from sweet potatoes. Most crops in Carver's time were only used for one particular purpose and that severely limited the economic growth of many crops. He opened the door for modern biotechnological applications involving the commercial manufacturing of plant products. Carver showed that it was possible to make a variety of materials including beverages, cheese, cosmetics, dyes, flour, inks, soaps, and wood stains from crops. Many of the environmentally friendly soy inks used today were founded on Carver's studies. Carver did a majority of his research at Tuskegee University in Alabama. He died on January 5, 1943.

Erwin Chargaff

Born in Austria in 1905, Chargaff did his doctoral research in chemistry at the University of Vienna. He then studied bacteriology and public health at the University of Berlin and later worked as a research associate at the Pasteur Institute in Paris. Chargaff move to the United States after being offered a position at Columbia University in New York in 1935. At Columbia University, Chargaff used paper chromatography and ultraviolet spectroscopy to help explain the chemical nature of the DNA structure. He showed that the number of adenine units in DNA was equal to the number of thymine and the number of units of guanine was equal to the number of cytosine. These findings provided the major clue that Francis Crick and James Watson needed to determine the double helix structure of DNA. His principle of DNA structure became known as Chargaff's Rule. Much of his later research focused on the metabolism of lipids and proteins. Starting in the 1950s, Chargaff starting making philosophical comments criticizing the scientific community. One of his famous quotes was, "Science is wonderfully equipped to answer the question 'How?' but it gets terribly confused when you ask the question 'Why?'" Chargaff died in New York in 2002.

Martha Chase

Martha Chase was born in Cleveland Heights, Ohio, in 1930. She was one of the few scientists to perform world-renowned research as an undergraduate student. Chase obtained her bachelor's degree in biology from the University of Dayton. A summer internship in Albert Hershey's laboratory at Carnegie Institution of Washington brought her in contact with DNA research. At Carnegie Institution, Chase helped in carrying out a famous experiment now known as the Hershey–Chase or Blender Experiment. This experiment showed that viruses replicated using DNA. Their highly creative study helped to confirm the role of DNA as being the chemical of genetic inheritance. She was in her early twenties when this epic study was completed. Geneticist Waclaw Szybalski of the University of Wisconsin–Madison stated, "I had an impression that she did not realize what an important piece of work that she did, but I think that I convinced her that evening. Before, she was thinking that she was just an underpaid technician." Chase then worked at Cold Springs Harbor to work at first Oak Ridge National Laboratory. She later earned a PhD in microbial physiology at the University of Southern California. Unfortunately, Chase's promising scientific career ended prematurely when she developed a disease that caused severe memory loss. She died from complications of pneumonia in 2003.

Stanley Cohen

Born to Russian Jewish immigrant parents in Brooklyn in 1922, Cohen was raised to value intellectual achievement. His family was too poor to pay for his college education. Cohen's father did not make much money as a tailor and his mother was a housewife. So, he studied biology and chemistry at Brooklyn College that did not charge tuition fees from New York City residents at the time he attended. Cohen then pursued a masters degree in zoology at Oberlin College in Ohio and a PhD in biochemistry at the University of Michigan. He financed his education with fellowships and by working as a bacteriologist at a milk processing company. His PhD research on the regulation of metabolism predated many of the genetic principles needed to fully understand the control of genes. Cohen took a position at Vanderbilt University in 1959 where he studied chemistry and biology of cell growth. His research led to the discovery of chemicals involved in skin growth and cancer cell development. As a result of his research, he was offered a research position with the American Cancer Society in 1976. In 1986 Cohen shared a Nobel Prize in Physiology or Medicine with Rita Levi-Montalcini of the Institute of Cell Biology in Rome, Italy. They received the award for their discoveries of growth factors essential for carrying out the cell culture techniques commonly used in biotechnology.

Stanley N. Cohen

Stanley N. Cohen was born in Perth Amboy, New Jersey, in 1935. He wanted to be a scientist while a young boy and showed an early interest in atomic physics. However, a high school biology teacher motivated Cohen to study genetics. Cohen studied biology at Rutgers University in New Jersey and obtained a medical degree from the University of Pennsylvania. He then accepted the job of a physician and a medical researcher at Stanford University in 1968. Stanford at time was a major research center for bacterial genetics. Consequently, Cohen developed a research interest in bacterial genetics and investigated the way bacteria acquire antibiotic resistance. He worked with Herbert Boyer to discover the methods used today for genetic engineering. Cohen's research helped Boyer produce the first genetically engineered products for the biotechnology company. Currently, Cohen is a professor of genetics and medicine at Stanford University. His research investigates cell growth and development. Cohen received many national awards and honors for his genetics research and medical studies.

Francis S. Collins

Francis Collins grew up on a small farm in the Shenandoah Valley of Virginia in the 1950s. His parents were highly educated people who believed in hard work and home schooling. Collins worked on the farm while doing the challenging home studies designed by his parents. He graduated high school at the age of 16 and went on to study chemistry and physics at the University of Virginia. Collins claims that he did not like biology because it was not as predictable as chemistry and physics. It was during his doctoral work at Yale that he developed an interest in genetics. He then wanted to use his knowledge of science for curing diseases. To achieve this new career goal he went on to complete a medical degree at the University of North Carolina. Collin's used his extensive training as a professor at the University of Michigan to identify the location of various genes that cause human disease. In 1989 his research team identified the gene for the debilitating genetic disorder cystic fibrosis. He also found the gene for Huntington's disease. In 1993, Collins was asked to be director of the National Center for Human Genome Research at the National Institutes of Health in Bethesda, Maryland. He continues to make contributions to biotechnology through his research in human genetics.

Gerty and Carl Cori

Gerty Theresa Cori was born Gerty Theresa Radnitz to a Jewish family in Prague, Czech Republic, in 1896. Carl Ferdinand Cori was also born in Prague, Czech Republic, in 1896. Gerty Cori was educated at home before entering a school for girls in 1906. She then attended the Medical School of the German University of Prague where Gerty Cori received an MD degree. Carl Cori's father, Dr. Carl I. Cori, was director of the Marine Biological Station in Trieste, Czech Republic. This gave Carl Cori an early interest in science. In 1914 he entered the German University of Prague to study medicine. Carl Cori served as a lieutenant in the Austrian Army during World War I. He then returned to complete his medical studies with his future wife, Gerty. Carl Cori held several research positions in Europe. The Coris immigrated to the United States when Carl Cori was offered a position at the State Institute for the Study of Malignant Diseases in Buffalo, New York. They then moved to the Washington University School of Medicine in St. Louis, Missouri, where both were offered research positions. The Coris studied metabolic diseases caused my abnormalities in sugar metabolism. Gerty Cori became a full professor in the same year she received the Nobel Prize in Medicine or Physiology with Carl Cori and Bernardo Alberto Houssay of Argentina. They received the award in 1947 for their research on metabolic diseases. Gerty Cori was the first American woman to win the Nobel Prize for Physiology or Medicine. Even today the basis of her research assists with new medical applications of biotechnology. Cori received many national honors and awards throughout her life. She died in 1957.

Francis Crick

Francis Harry Compton Crick was born in Northampton, England, in 1916. Although he is most known for his contributions to biology, Crick's primary interests were in physics. He studied physics during his undergraduate studies at University College in London. Crick then went on to do doctoral work in physics at the same university. The outbreak of World War II caused Crick to work as a military physicist for British Admiralty. After the war he went to Cambridge University in England to pursue graduate studies in biology. Crick worked in the molecular biology laboratory of Max Ferdinand Perutz where he was introduced to genetic research. Crick's previous work in X-ray crystallography paired him with the investigations of DNA structure being carried out by James Watson, Rosalind Franklin, and Maurice Wilkins. Their research on DNA structure grew out of their interest in the manner genetic information is stored in molecular form. Using X-ray crystallography data and cut-out paper models they hypothesized the double helix model of DNA structure. They published their results in a letter to the British jounal *Nature* in 1953. The name of the famous article is titled "Molecular structure of nucleic acids." This model of DNA structure proposed in the article was the hallmark study that spurred the growth of modern molecular genetics. In 1962, Crick was awarded the Nobel Prize in Physiology or Medicine that he shared with James Watson and Maurice Wilkins. Later in his career, Crick collaborated with Sydney Brenner investigating the biochemistry of protein synthesis. Crick died in San Diego, California, in 2004.

Charles Darwin

Charles Robert Darwin was born in 1809 in Shrewsbury, England. Darwin was raised in affluence and grew up with Unitarian values. He was destined to become a physician like his father, but was uncomfortable watching surgeries. In college he became active in naturalist societies and yearned to travel the world observing nature. He then began studying animal diversity with some of the greatest biologists in England. His father was unhappy with Darwin's interest in being a naturalist. It was not considered a noble profession for his family. Hence, Darwin's father enrolled him in college to become a minister. Darwin blended his theological education with his interest in nature to explore new ways of explaining animal and plant diversity. He developed a keen curiousity in geology and became frustrated by inconsistencies in the explanations of geological formations provided by opposing scientific writings. This spurred him to apply for a job as a naturalist on the HMS Beagle. It was from his observations on the Beagle that Darwin formulated the principles of evolution. Darwin is most noted for promoting the principles of natural selection. However, he unknowingly contributed to the mindset needed to develop biotechnology innovations. Darwin's observations about the natural selection of traits are still used by scientists to produce genetically modified crops with useful growing characteristics.

Félix d'Herelle

Felix d'Herelle was born in Montreal, Quebec, Canada, in 1873. He came from a French emigrant family and lost his father at the age of 6. D'Herelle's mother then moved the family back to France. His family had no resources to provide d'Herelle with a formal education. However, this did not stop him from pursuing an interest in microbiology. D'Herelle returned to Canada to set up a microbiology laboratory in his home. He taught microbiology to himself by reading scientific books and conducting experiments in his laboratory. At first, d'Herelle supported his family and his research by studying fermentation of foods for the Canadian government. He then held a variety of other jobs throughout the world requiring scientific expertise in spite of his lack of education. In 1910, while working in Mexico, he was investigating a disease that caused diarrhea and death in grasshoppers. The disease, it turned out, was caused by a bacterium in the intestines of the grasshoppers. He later went on to use the bacterium as a method of controlling the grasshoppers that caused significant crop loss. This strategy of biological control is still a biotechnology application in agriculture. D'Herelle then moved his family to Paris to work in the Pasteur Institute. At the Pasteur Institute, d'Herelle made his most notable discovery in 1915. He discovered the bacteriophage virus that attacks bacteria. Bacteriophages are important research tools in biotechnology and genetics. They played an important role in the discovery of DNA. Frederick Twort, an English biochemist, discovered the bacteriophage during the same year. So, both researchers are given credit for its discovery. D'Herelle continued to make many scientific and medical contributions until his death in 1949. Many scientists criticized d'Herelle for his lack of education. However, this did not stop the French Academy of Science from recognizing d'Herelle's long-lasting contributions to science.

Max Delbrück

Max Henning Delbrück was born in Berlin, Germany, in 1906. His father was a professor of history at the University of Berlin and his mother came from a professional family. So, Delbrück was expected to pursue a higher education. As a boy he was interested in astronomy and at first pursued an education in astrophysics. Delbrück then changed his research emphasis to theoretical physics in graduate school. He then directed his interests to chemistry after learning about the new research investigating atomic structure. This then led to a curiosity in biochemistry. In 1937, Delbrück took a position at the California Institute of Technology to study the growing field of fruit fly genetics. His move to the United States saved his life because most of his family was killed because of their resistance to the Nazi Party. Delbrück collaborated with Salvador Luria in 1942 to study the way bacteria are able to resist viral attack. This paved the way for understanding the benefitial nature of certain mutations. Delbrück was awarded the 1969 Nobel Prize in Physiology or Medicine for his discoveries on the stages of viral replication. He shared the prize with Alfred Hershey and Salvador Luria. Delbrück made another change in his research interests and began studying physiology. He is also noted for helping build one of the first molecular biology centers in Germany at the University of Cologne. Delbrück died in 1981.

Hugo de Vries

Hugo Marie de Vries was born in 1848 in the Netherlands. He studied botany at the University of Leiden in the Netherlands and completed his graduate studies at Heidelberg and Wurzburg Universities in Germany. De Vries returned to the Netherlands to become a professor of botany at the University of Amsterdam. At the university he performed plant breeding patterns that provided much insight into genetic variation. From his research he proposed the idea of genetic change through mutation long before anything was known about DNA. He published his findings about genetic change in a book called *The Mutation Theory* that was completed in 1903. De Vries also published supporting Darwin's hypothesis of pangenesis that describes the inheritance of characteristics. He is most noted for discovering a forgotten manuscript published by Gregor Mendel in the 1850s. Mendel's work provided de Vries with the information he needed to better understand the patterns of trait inheritance. De Vries then conducted experiments related to Mendel's original studies and published the results of his experiments in the journal of the French Academy of Sciences in 1900. A controversy was created when de Vries failed to reference the works of Mendel. This oversight was corrected and de Vries was credited with building the foundation for understanding inheritance patterns fundamental to biotechnology developments in agriculture and medicine. De Vries died in the Netherlands in 1935.

Renato Dulbecco

Renato Dulbecco was born in Catanzaro, Italy, in 1914. He developed an interest in physics while in high school. As part of a school science project, Dulbecco built a fully working electronic seismograph. He graduated from high school at the age of 16 and entered the University or Torino in Italy. Although he was interested in math and physics, Dulbecco decided to pursue medicine. He made this decision because he was fascinated by the work of an uncle who was a surgeon. At the University of Torino, he met two students who also went on to become famous scientists, Salvador Luria and Rita Levi-Montalcini. Dulbecco then went on the get his medical degree with a research interest in pathology. After medical school he joined the Italian Resistance movement against

Benito Mussolini during World War II. Dulbecco then left for the United States after the War to work with Salvador Luria at the University of Indiana. Dulbecco studied human viral diseases while at the University of Indiana. His research caught the interest of Max Delbrück. Delbrück asked Dulbecco to join him at the California Institute of Technology in 1949. In 1962, Dulbecco moved to the Salk Institute in California to perform genetic research on cancer. Dulbecco made a great medical study when he discovered that tumor viruses cause cancer by inserting their own genes into the chromosomes of infected cells. For this work he shared the 1975 Nobel Prize for Physiology or Medicine with David Baltimore and Howard Temin. Dulbecco continued doing cancer research helping with the advancement of biotechnology techniques for identifying and treating cancer. He was one of the major supporters of the Human Genome Project during its implementation. Dulbecco plans to continue doing research even past his 92nd birthday.

Paul Ehrlich

Paul Ehrlich was born into a Jewish family in Strehlen, Germany, (now in Poland) in 1854. Ehrlich's interest in science began early in his life when he would spend time learning to make microscope slides. He did undergraduate and graduate studies in biology. In addition, he earned a medical degree at the University of Leipzig in 1878. Ehrlich researched his interest in making microscope slides and developed many of the stains used today for studying cells under the microscope. He then went on to become a professor at the Berlin Medical Clinic where he continued his research on staining cells. Ehrlich then got involved in researching disease when he become director of a new infectious diseases institute set up at the clinic. He then started researching chemicals for controlling many devastating infectious diseases of humans. In 1908, Ehrlich shared the Nobel Prize in Physiology or Medicine with Ilja Iljitsch Metschnikow. Ehrlich received many national and international honors for his various research studies. He is noted for many discoveries that built the foundation for modern biotechnology. He is noted for his work in hematology, immunology, and chemotherapy. Ehrlich is noted for coining the term chemotherapy, which today is a common treatment for cancer and certain infectious diseases. Ehrlich was honored in Germany by having the street located by the Royal Institute of Experimental Therapy named Paul Ehrlichstrasse. During World War II the Nazi regime had the name removed because of Ehrlich's Jewish ancestry. However, after the War, his birth-place, Strehlen, was renamed Ehrlichstadt, in Ehrlich's honor. Ehrlich's methology for producing drug treatments and vaccines is a major contribution to modern biotechnology. He died in Germany in 1915 from a stroke.

Alexander Fleming

Alexander Fleming was born in 1881 in Lochfield, Scotland. He left the farming community to study medicine at St Mary's Hospital medical school in London. His medical experience as a captain in the Army Medical Corps spanned World War I where he became acutely aware of infections caused by battlefield wounds. This experience compelled Fleming to investigate the development of better antiseptics for reducing wound infections. Fleming returned to St. Mary's where he became a professor of bacteriology. In 1921, Fleming discovered a natural antiseptic chemical called lysozyme in tears and other body fluids. He then used the lysozyme as a standard for testing the effectiveness of other antiseptic chemicals he was researching. Some accounts claim that Fleming's lab was usually kept in disarray. This habit proved beneficial when Fleming discovered a fungus accidentally growing in a culture of bacteria. He noticed that the fungus reduced the growth of the bacteria. Fleming then referenced the research of Joseph Lister who in 1871 noticed that certain fungi inhibited the growth of bacteria. In 1928, Fleming made a similar observation and isolated the antiseptic chemical, which he named penicillin, from the fungus. Fleming was aware that he discovered a very powerful type of antiseptic that is today called an antibiotic. For this discovery, Fleming was awarded the Nobel Prize in Physiology or Medicine in 1945. He continued to investigate ways to battle disease including chemotherapy agents used for treating cancer. Many of his ideas are used to develop biotechnology drugs and medical treatments. He received many other awards for his research achievements. Fleming died in 1955.

Rosalind Franklin

Rosalind Elsie Franklin was born in London, England, in 1920. Franklin developed a keen interest in science as a young child. She was lucky to be at one of the few schools for women that taught chemistry and physics. Franklin's father was at first not supportive of her decision to study science in college. Her father did not believe that women should seek a higher education and wanted her to be a social worker. In spite of her father's wishes, she entered Newnham College where she studied chemistry and physics. Before completing her graduate studies she worked for the British Coal Utilization Research Association investigating the structure of carbon compounds. Franklin used the skills she learned at her job to carry out her doctorate studies in physical chemistry at Cambridge University. Upon finishing college she worked in Paris and then took a research position at King's College in London. It was at King's College she was asked to perform X-ray crystallography on DNA. Her experience at the British Coal Utilization Research Association gave her the expertise to analyze the physical properties of large organic molecules such as DNA. Her images of DNA structure helped Francis Crick, James Watson, and Maurice Wilkins in proposing the double helix structure of DNA. Franklin found it disturbing that her research was not published alongside the Watson and Crick article in the journal *Nature*. She left King's College to pursue a series of successful research on viral structure at Birkbeck College in London. Franklin continued doing research until developing cancer in 1956. She died in London in 1958. Many people felt she should have been honored along with Crick, Watson, and Wilkins for the 1962 Nobel Prize in Physiology or Medicine. However, she died before the award was given. At that time, the prize was awarded only to people who were alive when their achievement was recognized.

Galen

Galen was born Claudius Galenus of Pergamum in AD 131 in Bergama, Turkey. His father was a wealthy architect who valued education. As a child, Galen was fascinated by agriculture, architecture, astronomy, and philosophy. However, he concentrated his studies on medicine and trained to be physician who treated injured gladiators. He studied medicine in Greece and spent much of his life studying anatomy and physiology in Rome. What does an ancient physician have to do with developments in biotechnology? Biotechnology was based on many of the agricultural and scientific principles practiced in by early cultures. Galen set stage for a developing more rational approach to scientific methodology. Much of what was known about science in his society was based on untested hypotheses and philosophical arguments. His curiosity about the human body coaxed him to perform a variety of experiments on animals and injured gladiators. Many of the experiments he conducted on live animals would be considered cruel today. Galen made many human anatomical illustrations that were useful hundreds of years later. He also developed many types of surgical instruments and learned how to successfully carry out a variety to different surgical procedures. Galen found evidence against the accepted belief that the mind was in the heart and not the brain as Aristotle conjectured. His greatest contribution to biotechnology was instilling an awareness of the procedures needed to perform detailed studies of human health. Galen's strategy of doing science was the foundation for the modern scientific method. It is believed that he died between AD 201 and 216.

Archibald Garrod

Archibald Edward Garrod was born in 1857 in London, England. Having a father who was a physician, Garrod developed an early interest in biology. He first obtained a biology degree and then studied medicine at Oxford University. Garrod pursued graduate studies in medicine in Vienna, Austria. His interest in medicine focused on the factors that caused genetic diseases. During his time genetic errors were referred to as inborn diseases. This distinguished these conditions from infectious diseases known to be caused by microorganisms. Garrod was formulating the origins of genetic disorders before people understood the mechanisms of inheritance. He approached his research with the hypothesis that inborn diseases were due to errors of metabolism. Garrod presented this idea to the scientific community in his book *Inborn Errors of Metabolism* written in 1923. His research in graduate school led to his belief that inborn diseases were the result of altered or missing steps in the chemical pathways that made up metabolism. He studied several genetic disorders including albinism, alkaptonuria, cystinuria, and pentosuria. Albinism is due to the lack of a protein that affects eye, hair, and skin color. Alkaptonuria, cystinuria, and pentosuria are metabolic diseases that can be measured by chemical changes to the urine. Garrod's insights about genetic disorders are still the basis of understanding disease. It is the rationale for many medicines and for gene therapy. He received many national awards for his scientific findings. Garrod died in Cambridge, England, in 1936.

Walter Gilbert

Walter Gilbert was born to a well-respected professional family in Boston, Massachusetts, in 1932. His mother was a child psychologist and father was an economics professor at Harvard University. In an interview, Gilbert explained that he was educated at home by his mother who routinely gave him intelligence tests to measure his learning. His family then moved to Washington, DC, where he developed an interest in science while in high school. Gilbert returned to Massachusetts to study chemistry and physics at Harvard University. He then went to Cambridge University in England for his graduate studies where he met James Watson. His conversations with Watson spurred his interest in understanding the structure of RNA. RNA is the molecule that assists with the function of DNA. Gilbert was asked to take a professorship at Harvard where he became the first person to thoroughly explore the way RNA is involved in the synthesis of proteins. He made a variety of discoveries that provided a fundamental understanding of how genes carry out their functions. Other contributions to biotechnology include a rapid way to sequence the vast amount of information stored in the DNA's structure. He also paved the way for the genetic engineering of bacteria that produce medical compounds. For his work on gene function, Gilbert was awarded the 1980 Nobel Prize in Chemistry with Paul Berg and Frederick Sanger. He has received many other national awards and recognitions.

Frederick Griffith

Griffith was born in England in 1881. He studied medicine and became an army medical officer consigned to work on a vaccine against bacteria that caused pneumonia. While working with the bacteria he formulated the first hypothesis about the chemistry of inheritance. Before his discovery, scientists had little knowledge about the way traits were passed on from one generation to the next. While developing the vaccine, Griffith discovered two types or strains of the bacterium associated with pneumonia. One type he called the S strain because it had a smooth appearance in culture. The other type had a rough appearance. To make the vaccine he had to inject mice with the live bacteria to evaluate the immune response used to combat the bacteria. Griffith discovered that only the S strain of bacteria caused pneumonia. The R strain appeared harmless. Next, he injected killed S strain bacteria into the mice. This was done in order to isolate immune response chemicals harming the mice with the pneumonia disease. Then, for some unknown reason, Griffith injected the mice with a mixture of live R strain bacteria with S strain. It was assumed he was hoping to get a more powerful vaccine by doing this. To his surprise the mice died from pneumonia. Upon examining the mice he discovered live S strain bacteria in the mice. From this data he surmised that a chemical associated with the traits of the bacteria, now called genetic material, was transferred from the dead to the live bacteria. This research paved the way for further investigations into the chemistry of genetic material. Griffith died in 1941 before he was able to see a resolution to the debate about the chemistry of genetic information.

Henry Harris

Harris was born in Australia to a Russian immigrant family in 1924. At first he had little intent of becoming a scientist. Harris studied language in college and then developed a curiosity for medicine. He followed up on his new interest by receiving a medical degree from the Royal Prince Alfred Hospital in Sydney, Australia. Harris preferred doing medical research and then moved to England to study pathology at Oxford University. His research interest was in distinguishing the differences between normal cells and cancerous cells. Harris' most notable research involved the fusion of normal cells to cancer cells producing a cell called a hybridoma in 1969. This was a feat that was considered impossible by most biologists at that time. By doing this, he discovered a group of genes that shut down the cancerous properties of the cancer cells. This study provided the foundation for modern cancer research. It lead to the development of many biotechnology drugs that control cell growth. Hybridomas also became a biotechnology tool for producing vaccines and other medically important chemicals. Harris received recognition from The Royal Society in England for his achievements. In 2000, Harris authored a book called *The Birth of the Cell* highlighting the major achievements in cell biology. Harris of often referred to as one of the world's leading cell biologists.

Alfred Hershey

Alfred Day Hershey was born in Owosso, Michigan, in 1908. Hershey pursued a passion for science studying chemistry at Michigan State College. He then changed his interest to biology and completed a PhD in bacteriology at Washington University in St. Louis, Missouri. Upon graduation he accepted a position in the school of medicine at Washington University. In the 1940s, he began doing research on bacteriophage viruses with noted biologists Max Delbrück and Salvador Luria. The collaboration was formed because Delbrück was intrigued by Hershey's research findings. Delbrück felt it would be more productive if they combined their efforts to work out the mechanism of bacteriophage reproduction. Hershey then joined the research staff of Cold Spring Harbor in New York in 1950. Two years later he was joined by Martha Chase who helped him investigate viral reproduction using bacteriophages. Hershey and Chase developed on the famous Blender Experiment that showed how viruses replicated using DNA. This study confirmed the role of DNA as being the chemical of genetic inheritance. Hershey was awarded many honorary awards and degrees for his research efforts. In 1969, Hershey was awarded the Nobel Prize in Physiology or Medicine that he shared with Luria and Delbrück for their discovery of viral genetic sturture and replication. He is remembered as a competent researcher who was reserved in social settings. A colleague,

Franklin W. Stahl, described Hershey by the statement, "His economy of speech was greater even than his economy of writing. If we asked him a question in a social gathering, we could usually get an answer like 'yes' or 'no.'" Hershey died in 1997.

David Ho

David Ho was born in 1952 in Tai Chung, on the island of Taiwan. His original name was Ho Da-i which the family changed when they settled in America. Ho did not speak English when he arrived in America. He overcame his language barrier and went on to study physics at the Massachusetts Institute of Technology and the California Institute of Technology. Ho then changed his acadmic direction and obtained a medical degree from the Harvard Medical School in 1978. He returned to California to do residency training in infectious diseases at the Univeristy of California at Los Angeles School of Medicine. Ho was fornunate to work with some of the first recorded cases of AIDS. The severe nature of the disease compelled Ho to persue research in finding a treatment of AIDS. Ho's research cleared up many of the scientific misconceptions about AIDS virus reproduction. He also learned about the way the body's immune system failed during an AIDS infection. Ho developed the therapy called protease inhibitors and other drugs currently used to treat AIDS. His experimental approach in developing these treatments became a standard method used today in biotechnology drug applications. Ho is currently searching for a vaccine that will hopefully wipe out the deadly outcomes of AIDS.

Leroy Hood

Leroy Hood was born in Missoula, Montana, in 1938. In an interview he said that he credits his success to the very high standards of excellence that his parents expected in school and in all other chosen endeavors. His parents instilled the values of independent thinking in Hood while he was a child. In high school, Hood was involved in many academic pursuits and became a student leader in academics, sports, and student government. Hood entered the California Institute of Technology where he was exposed to the renowned scientists on the faculty. Their depth of knowledge and enthusiasm compelled Hood to study the sciences. Hood then earned a medical degree from Johns Hopkins University in Maryland and a PhD in biochemistry from the California Institute of Technology. His first research position was at the California Institute of Technology. Hood then became a professor in the immunology department at the University of Washington,

School of Medicine. Most of his research focused on the development of procedures for identifying genetic diseases. Many of his discoveries are fundamental to biotechnology applications used in treating genetic disorders. Currently, Hood is president and the co-founder of the Institute for Systems Biology in Seattle, Washington. Hood is recognized as one of the world's leading scientists in molecular biotechnology and genomics. He founded many biotechnology companies, including Amgen, Applied Biosystems, Darwin, MacroGenics Rosetta, and Systemix.

Robert Hooke

Robert Hooke was born in 1635 on the Isle of Wight south of England. He was educated at home by his father John Hooke who was in the clergy and served as Dean of Gloucester Cathedral. Hooke planned to be an artist and even did an art apprenticeship before college. However, he developed an interest in science at Oxford University after working with some of the great British scientists of that era. After working in various scientific jobs, Hooke became a professor of geometry at Gresham College in London. He made a variety of scientific contributions mostly in the fields of architecture, mathematics, and physics. However, he is most noted for his contribution to the biological sciences. Hooke became famous in the public and the scientific community with the publication of his book *Micrographia*, published in 1665. Hooke's book contained the first microscopic images of cells and minute animals. This fascinated the scientific community and paved the way for scientific investigations using the microscope. A noted scholar and member of Parliament, Samuel Pepys, wrote the following comment about Hooke's book, "Before I went to bed I sat up till two o'clock in my chamber reading Mr Hooke's Microscopical Observations, the most ingenious book that ever I read in my life." The microscopic examination of cells remains a critical component of modern biotechnology. Hooke was considered the single greatest experimental scientist of his century. His writings show that he was deeply knowledgeable about diverse sciences and technologies such as architecture, astronomy, biology, chemistry, geology, naval technology, and physics. He died in London in 1703.

John Hunter

John Hunter was born in 1728 in Long Calderwood, Scotland. He studied anatomy and surgery at St. Bartholomew's Hospital in London. Hunter then became an instructor of anatomy and surgery at

St. George's University of London. He also was a British army surgeon where he researched and treated infections associated with gunshot wounds and other injuries. Hunter is most noted for elevating the practice of surgery from what was considered a "technical trade" to a medical science. During his medical training, Hunter was appalled by the lack of science that went into surgical practices. Like many of the other earliest contributors to biotechnology, Hunter rejected the argumentation and speculation that dominated medical thinking. He insisted on experimentation and direct observation when studying disease and injury. The rationale for all biotechnology cures and treatments are founded in the ideology promoted by Hunter. His research contributions include investigations into the inflammatory process and sexually transmitted diseases. Hunter is considered one of the three greatest surgeons of all time because of his keen attention to detail and his "Don't think, try" attitude. His legacy is honored by John Hunter Hospital in Sydney, Australia. The hospital was named after three John Hunters who contributed to the development of Australia. Hunter died in London, England, in 1793.

François Jacob

François Jacob was born in June 1920 in Nancy, France. He had an early interest in medicine and pursued a medical degree at the University of Paris. However, his medical education was interrupted by the German invasion of France. Jacob escaped to England where he joined the Free French forces and fought the German forces in Normandy, France, and North Africa. After the War, Jacob returned to the University of Paris to finish his medical degree. He decided not to practice medicine because of physical limitations from war injuries. This decision compelled him to complete doctoral studies in biology so he could do research. Jacob did most of his research at the Pasteur Institute in Paris where he worked with geneticist André Lwoff. Most of Jacob's research focused on the function of bacterial and viral genes. His discoveries complemented the findings of Jacques Monod. Together, their research provided the model for gene function that was essential for the growth of biotechnology. Their theory is the basis of controlling the traits of genetically modified organisms. Jacob shared the 1965 Nobel Prize in Physiology or Medicine with André Lwoff and Jacques Monod for their research on the genetic control of protein synthesis. He was awarded numerous national awards for his scientific achievements. Jacob changed his research emphasis to molecular evolution and published a book on this topic and other aspects of genetics.

Zacharias Janssen

Zacharias Janssen was born in 1580 in Middleburg, Holland. His inquisitive mind as a child gave him an interest in the science of optics. This curiosity was fostered by his father Hans who designed spectacle lens. At 15 years of age, it is believed that Janssen and his father invented the forerunner of the modern compound microscope. Some historians believe that his father built the first one, but young Janssen produced others for sale. Janssen's microscope consisted of two tubes that slid within one another and had a lens at each end. The microscope was focused by sliding the tubes until the specimen was seen as a clear image. It was not a powerful microscope and only magnified a specimen three to nine times its size. Magnification was adjusted by changing the size of a covering called a diaphragm placed over the lens closer to the specimen. This early microscope was more of a curiosity than a scientific tool. Its low magnification provided little ability to study biological samples. However, it motivated other lensmakers to build more powerful microscopes for scientific purposes. Biotechnology would not have become a science if it were not for people like Janssen who created the tools for investigating the structure of living organisms. Janssen worked as a lensmaker like his father and died in 1638.

Alec Jeffreys

Sir Alec John Jeffreys was born in 1950 in Luton, England. Jeffreys was interested in biology and chemistry as a child. He was known for carrying out many experiments around the house. A microscope as gift kept him occupied throughout elementary school. Jeffreys went on to study molecular biology at Oxford University in England. He then took an academic position at the University of Leicester after receiving his PhD at Oxford. In 1984, a chance discovery of highly variable regions of DNA gave him the idea to develop a technique called DNA fingerprinting. At the time of his discovery Jeffreys said he had a "eureka moment" explaining, "I thought—My God what have we got here but it was so blindingly obvious. We had been looking for good genetic markers for basic genetic analysis and had stumbled on a way of establishing a human's genetic identification. By the afternoon we had named our discovery DNA fingerprinting." DNA fingerprinting became a popular tool of biotechnology immediately after Jeffreys published his findings. His technique became the standard way of identifying DNA for a variety of purposes. DNA fingerprinting made national news when it was used to identify the rapist and killer of two girls in Leicestershire, England, in 1983 and 1986. Jeffreys maintains an interest in unusual sequences of DNA. He is involved in a variety of research projects including studies on the evolution of genes. Jeffreys has been honored with many awards for his contributions to biotechnology.

Edward Jenner

Edward Jenner was born in 1749 in the small village near Gloucestershire, England. He showed an early interest in science and as a young man he wrote observations about nature that were previously not recorded in the scientific literature. Jenner went to school in Wotton-under-Edge and Cirencester where he completed his medical training. His rural upbringing exposed Jenner to a variety folklore about healing and medical remedies. He began testing the validity of the some of the tales in his home laboratory that he privately funded. One story that he tested became the basis of modern vaccinations. Jenner investigated the story that milkmaids did not develop the devastating viral disease called smallpox. He developed a hypothesis that an old practice called variolation would be effective at preventing smallpox. However, Jenner's variolation differed from the usual practice developed in Asia. Variolation traditionally involved scratching a person with infected fluids to produce protection against the particular disease. Doing this with smallpox would have been dangerous. Jenner proposed doing variolation against smallpox using the pus of milkmaids exposed to a related cattle disease called cowpox. He believed that exposure to the cowpox prevented milkmaids from getting smallpox. Jenner tested his hypothesis in spite of much resistance from society and the medical community. His technique worked and provided a safe way of ridding Europe of smallpox. Jenner encountered the prejudices of the established medicals that dominated London at the time. His findings were not taken seriously because he was considered an unsophisticated country doctor. His gift to biotechnology was the strategy for producing vaccines against infectious disease. Like many great scientists, Jenner's ability at enquiry was his ground-breaking contribution to the science of medical biotechnology. Jenner died in 1823 and was honored by having the Edward Jenner Institute for Vaccine Research in Compton, England, established in his name.

Ernest Everett Just

Just was born in 1883 in Charleston, South Carolina. His father died when Just was only 4 years old. As a child he had to work as a farm hand to help financially support the family. Just's mother, who was a teacher, sent him to high school in New Hampshire to avoid the poor educational opportunities for African Americans living in South Carolina at that time. Just showed his academic talents in college. He was the only person at Dartmouth College to graduate with honors in botany, history, sociology, and zoology. Upon graduation for Dartmouth College, he accepted a faculty position at Howard University in Washington, DC. He went to Howard because there were few college teaching opportunities for African Americans when he graduated in 1907. Just continued his education part-time at Woods Hole in Massachusetts and earned a PhD from the University of Chicago in 1916. His graduate work was in experimental embryology. Just's work at Woods Hole was awarded with the first Springarn Medal in 1915 for pioneering research on fertilization and cell division. He became a world-renowned expert in cell development and identified the importance of the cytoplasm in controlling cell development. Just was requested to give lectures around the world about his research on the cell membrane and cytoplasm. His work forms the foundation for the current strategies used in biotechnology laboratories performing stem cell research. Just died in Washington, DC, in 1941.

Har Gobind Khorana

Har Gobind Khorana was born in 1922 to poor Hindu parents in Raipur, Pakistan. His family was one of the few literate families in the area and his father insisted that the children pursue higher education. He attended Dayanand Anglo Vedic High School in Multan where he was influenced by one of his teachers to study science at Punjab University in Lahore, Pakistan. Khorana's excellence in college awarded him the opportunity to obtain a PhD at the University of Liverpool in England. During his studies he helped in discovering the way the four different types of nucleotides are arranged on the DNA to determine the chemical composition of a gene. Khorana discovered an important piece of the genetic code called the stop codon. It is the information that tells the cell where the information for a particular gene ends. Khorana shared the 1968 Nobel Prize in Physiology or Medicine with Robert W. Holley and Marshall W. Nirenberg for their interpretation of the genetic code and its function in protein synthesis. He was awarded fellowships and professor positions in Switzerland at the Swiss Federal Institute of Technology and the University of British Columbia in Canada and at the University of Wisconsin. In 1971, Khorana took a position at Massachusetts Institute of Technology. One of his most recent contributions to biotechnology was the synthesis of the first artificial copy of a yeast gene. This technology is a standard technique used in contemporary genetic engineering.

Shibasaburo Kitasato

Kitasato was born in Oguni on Kyushu Island, Japan, in 1852. He received a medical education at Kumamoto Medical School and Imperial University. Kitasato had been doing bacteriology research. This motivated him in 1885 to work with Robert Koch and Emil von Behring in Berlin, Germany. Germany at the time was a major center for bacterial disease research. Kitasato studied toxins produced by bacteria that cause anthrax, diphtheria, and tetanus. Anthrax is a cattle disease that causes severe internal bleeding in humans. Diphtheria is a serious throat infection that was a major cause of death in children during Kitasato's time. Tetanus is a potentially fatal disease of infected wounds. It causes paralysis that eventually stops a person from breathing. Kitasato's research led to the development of vaccines that block the effects of the bacterial toxins on the body. These vaccines are called antitoxins. Antitoxins have many valuable medical purposes including being used as antivenoms that protect against bites from venomous snakes. The theory behind antitoxin production became the basis for developing many types of medical diagnostic tests including the home pregnancy test. He is also known for co-discovering the bacterium that causes plague in 1894. Kitasato returned to Japan in 1891 and set up an institute for the study of infectious diseases. The institute was taken over by Tokyo University in 1914. Kitasato left Tokyo University to form the Kitasato Institute in 1918. Today, Kitasato Institute is involved in the production of new drugs and vaccines for fighting infectious diseases.

Robert Koch

Robert Koch was born in 1843 in Clausthal, Germany. Koch was one of 13 children. He showed incredible intellectual abilities at an early age by teaching himself to read newspapers at the age of 5. Koch was also known as a fan of classical literature and as a keen chess player. He developed an interest in science while in high school and intended on pursing biology in college. In 1866, Koch completed a medical degree from the University of Gottingen in Germany. While in medical school he developed a strong interest in pathology and infectious diseases. Koch served as a physician in several towns throughout Germany and then volunteered as a military surgeon during the Franco-Prussian war from 1870 to 1872. After his military service he became district medical officer for Wollstein, in what is now Poland. His major interest as a medical officer was tracing the spread of infectious bacterial diseases. He was particularly interested in the transmission of anthrax from cattle to humans. However, Koch was dissatisfied with the current methods of confirming the cause of infectious disease. By 1890, he meticulously developed four criteria that must be fulfilled in order to establish a cause of an infectious disease. These criteria are called Koch's postulates or Henle-Koch postulates. Friedrich Gustav Jacob Henle collaborated in Koch's research. In 1905, Koch's work was recognized by being awarded the Nobel Prize for Physiology or Medicine. The medical applications of biotechnology still rely on the Koch's principles of confirming the causes of infectious diseases. Koch died in 1910 in Black Forest region of Germany.

Arthur Kornberg

Arthur Kornberg was born in Brooklyn, New York, in 1918. His parents settled in New York after leaving Poland in 1900. He excelled academically in New York City public schools. Kornberg received an undergraduate degree at the City College of New York and then a medical degree from the University of Rochester in 1941. While in medical school Kornberg was noted for discovering the prevalence of a common but harmless genetic condition in the a liver called Gilbert syndrome. He surveyed his fellow students to discover how common the condition was. Kornberg also had Gilbert syndrome and published the results as his first research paper while doing an internship in internal medicine in 1942. After his internship, Kornberg became a ship's physician for the United States Navy and then worked at the National Institutes of Health, in Bethesda, Maryland, from 1942 to 1953. His performed research on the enzymes involved in the metabolism of nucleic acids. While at the National Institutes of Health, Kornberg did some training with Severo Ochoa at New York University School of Medicine and with Carl Cori at Washington University School of Medicine in St. Louis. Ochoa contributed to an understanding of the structure of DNA. He shared the 1959 Nobel Prize in Physiology or Medicine with Severo Ochoa for their work on the discovery of the mechanisms in the biological synthesis of DNA. Kornberg then took a position doing genetics research at the Stanford University School of Medicine where he set up the biochemistry department. In 1991, Kornberg started researching the evolution of DNA. During the creation of modern biotechnology, Kornberg caused some public controversy by commenting, "A scientist shouldn't be asked to judge the economic and moral value of his work. All we should ask the scientist to do is find the truth and then not keep it from anyone."

Philip Leder

Philip Leder was born in 1934 in Washington, DC. He showed keen intellect as a child and graduated from Harvard University with honors in 1956. Leder then went on to obtain a medical degree at Harvard Medical Center. He showed a great interest in doing medical research. This led him to pursue an internship at the National Institutes of Health in Bethesda, Maryland. In 1963, Leder started doing genetics research in Marshall Nirenberg's laboratory at the National Institutes of Health. While at Nirenberg's laboratory he helped devise a test called the triplet binding assay. This procedure greatly led to the understanding of the genetic code. The technique paved the way for many biotechnology developments that required information about the DNA sequence of particular genes. It also motivated researchers to find even faster techniques for interpreting genomic information. Leder left the National Institutes of Health to become chair of the Department of Genetics at Harvard Medical School. At Harvard, he contributed much to the understanding of many genes. In 1982, he developed a genetically altered "oncomouse" used to assist cancer research. His "oncomouse" became a model tool for future biotechnology developments. The "oncomouse" produced much controversy about the patenting of living organisms. Leder received many awards and honors for his research. He continues to do research on the genetics of cancer and embryological development at Harvard Medical School and at the Howard Hughes Medical Institute in Maryland.

Joshua Lederberg

Joshua Lederberg was born in Montclair, New Jersey, in 1925. His parents were recent immigrants from a region of Palestine, now known as Israel. They came to the United States to avoid the violence taking place where they lived. The family moved to New York City where he received his education in public schools. Lederberg excelled in school and showed an early interest in science. He studied zoology at Columbia College in New York and obtained a PhD in microbiology at Yale after graduating high school at the age of 15. Lederberg once commented that his success in school was driven by "an unswerving interest in science, as the means by which man could strive for an understanding of his origin, setting and purpose, and for power to forestall his natural fate of hunger, disease and death." Upon leaving Yale he was offered a professorship in genetics at the University of Wisconsin. He was only

22 years old at that time. Lederberg then moved to Stanford School of Medicine before going to Rockefeller University in 1978. He received the 1958 Nobel Prize in Physiology or Medicine with George Beadle and Edward Tatum for their discovery that genes act by regulating chemical events in the cell. Lederberg was called a prodigy because he received the prize when he was only 33 years old. He was also recognized for his discoveries concerning genetic recombination and the organization of the genetic material of bacteria. All of his research helped in paving the way for genetic engineering and many of the principles of modern biotechnology.

Antony van Leeuwenhoek

Antonie van Leeuwenhoek was born in 1632 in Delft, Netherlands. He is usually referred to as the "father of microbiology." His interest in studying the microscopic structure of life developed from his curiosity of microscopes which were invented by Zacharias Janssen around 1595. Leeuwenhoek never attended a university. With no formal scientific training, he improved upon the design of the microscope until he developed one capable of magnifying specimens upto 300 times their normal size. Using a razor and his homemade microscopes, Leeuwenhoek investigated the microscopic structure of animals and plants. He was the first person to see red blood cells and sperm. At that time scientists were not aware of the sperm and had only speculations about the passage of traits into offspring. Leeuwenhoek was also the first person to see bacteria collected from his teeth and from various samples of water collected from ponds and streets. His discovery of bacteria helped build the idea that infectious diseases were caused by microorganisms. Leeuwenhoek called the microorganisms "animalcules" meaning that they were small life forms. At first he worked without any contact with the scientific community. A friend encouraged Leeuwenhoek to communicate his findings with the Royal Society of England. Leeuwenhoek published over 300 letters describing his findings. His discoveries first met skepticism and much criticism because most scientists had contrary views to Leeuwenhoek's observations. His work persisted and formed the basis of many modern biological principles that are used in all biotechnology applications. Leeuwenhoek's curiosity heralded in a new way to investigate nature.

Rita Levi-Montalcini

Rita Levi-Montalcini was born with her twin sister Paola in Turin, Italy, in 1909. Her father was a mathematician who worked as an engineer and her mother was an artist. The parents valued education for all of the children. However, her father discouraged the women in the family from seeking professional careers because he believed it would interfere with their ability to take care of a family. Levi-Montalcini requested the permission of her father to seek an academic track in high school. Upon graduation she pursued a medical degree at the University of Turin. While studying medicine she did research on nerve cell growth. Because her family was Jewish, they were forced into hiding in Florence during World War II when the Germans occupied Italy. In spite of living in exile, Levi-Montalcini continued doing research in a secret laboratory in her home. Levi-Montalcini had to rebuild the laboratory in another location when her house was bombed when the American army was fighting the Germans. She studied the effects that amputating the limbs of chickens had on the nervous system development. In 1946, she moved to the United States to continue her research at Washington University in Saint Louis, Missouri. While at Washington University, she discovered proteins called growth factors that determine how the nervous system forms. For this research, Levi-Montalcini shared the 1986 Nobel Prize in Physiology or Medicine with Stanley Cohen. Growth factors are very important for many biotechnology applications requiring cell culture and stem cell research.

Salvador Luria

Salvador Edward Luria was born in Torino, Italy in 1912 and was of Jewish heritage. Luria had an early interest in science and attended the University of Turin to pursue a medical degree. He studied radiology while at medical school and then served as a medical officer in the Italian army. After leaving the army, he studied at the Physics Institute of the University of Rome. Luria had to leave Italy for France in 1936 because his socialist and anti-War philosophy was contrary to the fascist government of Italian dictator Benito Mussolini. While in France, Luria studied at the Curie Laboratory of the Institute of Radium in Paris. The Nazi invasion of Europe forced Luria to flee to the United States in 1940. Luria held various academic positions in the United States. He did research at Columbia University in New York City and then served as a professor at Indiana University, the University of Illinois, and the Massachusetts Institute of Technology. Luria ultimately became director of the Center for Cancer Research at Massachusetts Institute of Technology in 1974. Luria's early research on viral genetics uncovered the way that viruses reproduce. He then became one of the world's foremost virologists. Luria shared the 1969 Nobel Prize in Physiology or Medicine with Max Delbrück Alfred D. Hershey for their contributions to viral genetics. He died in Lexington, Massachusetts, in 1991.

André Lwoff

Andre Michael Lwoff was born in 1902 in Ainay-le-Château, France of Russian and Polish parents. Lwoff's intense interest in biology encouraged him to pursue a job at the Pasteur Institute while working on his medical studies at the University of Paris. His interest in research caused him to continue his studies at the University of Paris to pursue a PhD in infectious diseases. He was interested in how certain parasites caused diseases in various animals. Lwoff stayed on at the Pasteur Institute carrying out microbiology research. However, he also conducted research at the University of Heidelberg in Germany and at Cambridge University in England. His research focused on the way certain bacteria and viruses cause disease. His work on the virus that causes polio led to an understanding of viral disease that is applied in many contemporary biotechnology developments. Lwoff's most noted research was the discovery that certain viruses can insert their DNA into the DNA of bacteria that they infect. Subsequently, the bacteria pass the viral DNA on to succeeding generations of bacteria. He shared the 1965 Nobel Prize in Physiology or Medicine with François Jacob and Jacques Monod for his discovery that the genetic material of a virus can be incorporated into the DNA of bacteria. He received many other international awards and honors. Lwoff died in Paris in 1994.

Barbara McClintock

Barbara McClintock was born in Hartford, Connecticut, in 1902. She developed an early interest in science and entered Cornell University with the intent of studying biology. McClintock showed a curiosity in the newly forming field genetics and was invited by a professor to enroll in the only genetics course open to undergraduate students at the university. Women were not encouraged to major in genetics at Cornell at the time. In spite of this, she became part of a small group people who studied corn genetics at the cellular level. Her interests focused on the chromosome structure. She was quoted about her interest in genetics saying, "I have pursued it ever since and with as much pleasure over the years as I had experienced in my undergraduate days." McClintock performed this work through graduate school and obtained a PhD in botany from Cornell University in 1927. She taught at Cornell for a short time and then was awarded two fellowships related to her research. In 1936, McClintock became a professor at the University of Missouri, Columbia.

She then left Missouri to pursue her interest in molecular genetics at the Carnegie Institution of Washington, Cold Spring Harbor, New York. McClintock's research at the Carnegie Institution of Washington provided the honor of being the third woman elected to the prestigious National Academy of Sciences in 1944. During the 1940s and 1950s, McClintock showed how certain genes were responsible for turning on or off particular characteristics plants. Her greatest discovery was a movable gene called a transposable element. Its ability to relocate itself on the chromosome was contrary to the contemporary beliefs of biologists. Transposable elements were exploited for many biotechnology applications. McClintock was awarded the Nobel Prize in Physiology or Medicine 1983 for her discovery of transposable genetic elements. She received many other international and national awards for her work. McClintock died in 1992.

Ilya Mechnikov

Ilya Ilyich Mechnikov, also known as Eli Metchnikoff, was born in a village near Kharkoff in Russia, which is now in Ukraine, in 1845. He showed an early fascination for nature and was said to give lectures to other children about natural history. Mechnikov studied natural sciences at Kharkov University and completed his undergraduate degree in 2 years. He continued his education in Germany at the University of Giessen, the University of Göttingen, and the Munich Academy. Mechnikov's research in Germany focused on the digestive processes of invertebrates. He then performed doctoral studies on embryological development in Naples, Italy, before returning to Russia in 1867 to work at the newly formed University of Odessa. His research was interrupted by the loss of his two wives to disease and periods of depression that led to two unsuccessful suicide attempts. Mechnikov's studies in Odessa focused his research on the immune system and discovered important aspects of how the body fights disease. His findings formed the basis for the theory of immunity which is a common component of many biotechnology advances. Mechnikov shared the 1908 Nobel Prize in Physiology or Medicine with Paul Ehrlich for their research on immunity. He received many other awards including having Mechnikov University in Odessa, Ukraine, named in his honor. Mechnikov died in Paris in 1916 where he was doing research at the Pasteur Institute.

Gregor Mendel

Gregor Johann Mendel was born the son of a peasant farmer in 1822 in Heizendorf, Austria, which is now in the Czech Republic. Mendel began his studies at the St. Thomas Monastery of the Augustinian Order in Brünn in 1843. He was ordained into the priesthood in August of 1847 and immediately went to work as a pastor. From 1851 to 1853, Mendel studied botany, chemistry, physics, and zoology at the University of Vienna. He did this with the intent of teaching biology and mathematics. Mendel felt he would be better at research and teaching than at being a pastor. Unfortunately, he failed the teaching certification examination several times. Mendel returned to the monastery and was able to teach part-time aside from his other duties. It was at the monastery that Mendel did his pioneering work on the patterns of inheritance. Much of the early works on inheritance developed in the Middle East were destroyed. Mendel rediscovered the rules of heredity by observing the passage of traits in plants grown at the monastery. He conducted meticulous studies on pea plants observing how various characteristics of their flowers and seeds were passed on from one generation to the next. Mendel carried out his experimental work in the monastery garden for 8 years. His work culminated in a book titled *Treatises on Plant Hybrids* (Versuche über Pflanzen-Hybride). Unfortunately, the book was not noticed by contemporary scientists. Mendel gave a variety of lectures about his research findings. However, the audience did not comprehend the importance or significance of the work. Mendel's findings remained obscure until 1900 when the science was in place to better understand the mechanisms of inheritance. Now his work serves as the foundation of biotechnology. His intellectual achievement as a scientist was due to his ability at making knowledgeable hypotheses and accurate experiments from his observations. Mendel died in 1884 in Brno, Czech Republic, from kidney failure.

Johann Friedrich Miescher

Johannes Friedrich Miescher was born in 1844 in Basel, Switzerland. He came from a family of eminent scientists from Switzerland. His father, Johann Friedrich, and his uncle, Wilhelm His, were physicians who taught at the University of Basel. Meischer was also interested in medicine. However, he decided to study physiology in college believing that his partial deafness would impair his ability as a physician. He did not let his handicap stop him from seeking a career in science and from being a fan of classical music. Meischer pursued his college education in Germany studying organic chemistry at the University of Tübingen and physiology at the University of Leipzig. He returned to Switzerland to complete his doctorate in physiology at the University of Basel in 1868. Meischer went back to the University of Tübingen to study the chemistry of blood from the bandages of troops injured in the Battle of Crimea. He made a fundamental contribution to biotechnology by discovering nucleic acids in the nucleus of white blood cells. It was later learned by other scientists that the nucleic acid detected by Meischer was DNA. While working at the University of Basel, Meischer, isolated DNA from the sperm of salmon. This discovery gave him and other scientists the idea that DNA could be the inheritable material of an organism. Meischer made other important discoveries in physiology until he died Davos, Switzerland, in 1895.

César Milstein

César Milstein was born in 1927 in Bahía Blanca, Argentina. His father was a Jewish immigrant and his mother was a teacher from a poor family. The family had little money, but both parents saved enough funds so that all of the children could go to college. Milstein claimed to have been an average student. His involvement in student activities and politics kept his interest in school. He did his undergraduate studies at Colegio Nacional in Bahia Blanca and completed his PhD in biochemistry at the University of Buenos Aires in Argentina. Milstein then taught and did research at University of Buenos Aires until taking a leave of absence to work at Cambridge University in England. While at Cambridge he finished another PhD with a research emphasis in enzyme function. At Cambridge he was introduced to many of the pioneers of early genetics. He went back to Argentina but eventually settled in England to work at Cambridge University. Milstein shifted his research emphasis to immune system function. This is when he devised a technique for fusing white blood cells with tumor cells producing hybridoma cells or monoclonal antibodies. His discovery heralded in many medical applications of biotechnology. He shared the 1984 Nobel Prize in Physiology or Medicine with Niels K. Jerne and Georges J. F. Köhler for their research on the immune system and the discovery of monoclonal antibodies. Milstein received numerous international awards for his research. He died in Cambridge, England, in 2002 from a heart condition that he battled for many years.

Jacques Monod

Jacques Lucien Monod was born in Paris, France, in 1910. His father, a painter, and his mother, an American, moved the family to southern France where Monad spent most of his childhood. Monad's parents stressed the pursuit of cultural and intellectual activities. In addition, his father encouraged Monod to read Darwin and related writings. This motivated Jacob to study natural science at the University of Paris. He completed his undergraduate studies and PhD at the university. His contact with André Lwoff interested Monod in doing research on microorganisms. Monod received a grant to study microbial genetics at the California Institute of Technology. He returned to France to work at the Pasteur Institute. Monod ultimately ended up becoming director of the Cell Biochemistry Department at the institute. He also was a professor at the Collège de Sorbonne and the Collège de France in Paris. Monad shared the 1965 Nobel Prize in Physiology or Medicine with André Lwoff and François Jacob for their discoveries explaining gene function. His findings form the foundation of the genetic engineering principles. He was awarded many other honors including the medal of the French Legion of Honor. In addition, the Institut Jacques Monod in Paris was named in his honor. Monod died in 1976 in Cannes, France.

Thomas Morgan

Thomas Hunt Morgan was born in 1866 in Lexington, Kentucky. Hunt was very curious about nature while very young. He collected bird eggs and fossils around the area he grew up. As a result of this interest he obtained a bachelor's and a master's degree in biology from the University of Kentucky. Morgan performed some marine biology research in Massachusetts and in the Caribbean before going back to school to finish his graduate education. He then worked on his PhD at Johns Hopkins University in Baltimore, Maryland, where he studied animal development. Upon receiving his PhD, Morgan was offered a fellowship at the Marine Zoological Laboratory in Naples, Italy. Morgan changed his research emphasis to experimental embryology during the fellowship period. He took a position at Bryn Mawr College in Pennsylvania and then became a professor of experimental zoology at Columbia University in New York. Morgan's research findings in embryo development were contrary to many of the contemporary views of evolution and genetics. He began a series of experiments on fruit flies to investigate the role of chromosomes in passing along inherited traits. His experiments identified the hereditary units scientists now call genes. Morgan's groundbreaking research led to an invitation to develop the biology department at the California Institute of Technology. He received the 1933 Nobel Prize in Physiology or Medicine for his research about the role of chromosomes in heredity. Morgan received many international honors for his work in embryology and genetics. In addition, Morgan was honored by having the Thomas Hunt Morgan School of Biological

Sciences at the University of Kentucky named after him. Hunt died in 1945 in Pasadena, California.

Hermann Muller

Hermann Joseph Muller was born in 1890 in New York City. Muller's father encouraged in him an interest in the process of evolution and the scientific explanation of the origins of the universe. His early interest in science inspired Muller and two friends to form the first science club at Morris High School in Bronx, New York. Muller carried out his undergraduate and graduate studies in biology at Columbia University. He developed a strong interest in genetics after encountering the works of two notable genetics professors: Edmund Wilson, who discovered the cellular approach to heredity, and Thomas Morgan, who was the first to identify genes. After receiving his PhD, Muller had faculty positions at Rice Institute in Houston, Texas, and the University of Texas in Austin. In 1926, Muller confirmed that X-rays were responsible for causing mutations and other changes to chromosomes. These findings led him to oppose the overuse of X-rays for diagnosing and treating diseases. He campaigned for safety guidelines that ensure the protection of people who were regularly exposed to X-rays. Muller's socialist political views compelled him to work at the Institute of Genetics in Moscow, USSR. He remained there until 1937 when Soviet biological research became corrupted by political influences. Muller then worked at the Institute of Animal Genetics in Edinburgh, Scotland, and then returned to the United States to become professor of zoology at Indiana University. In 1946, Muller won the Nobel Prize in Physiology or Medicine for discovering of the role of X-rays in producing mutations. This finding became a valuable tool for producing novel genes used in biotechnology applications. Muller died in 1967 in Indianapolis, Indiana.

Kary Mullis

Kary Banks Mullis was born in Lenoir, North Carolina, in 1944. His parents were from rural farming backgrounds. Mullis claims to have spent many hours investigating the diversity of organisms living around the farmlands. He went to high school in Columbia, North Carolina, and then obtained a B.S. in chemistry from Georgia Institute of Technology in Atlanta. Mullis obtained a PhD in Biochemistry from the University of California at Berkeley. His research was on protein structure and synthesis. Mullis had broad scientific interests. He published in various disciplines and invented a variety of devices. Mullis did not seek an academic career. Rather, he applied his keen scientific mind as a scientist for the Cetus Corporation in Emeryville, California. Cetus was a biotechnology company established in Berkeley, California, in 1972. It was one of the companies that helped spur the growth of the biotechnology industry. While at Cetus, Mullis used his genetics training from the University of California at Berkeley to develop a procedure called the polymerase chain reaction based on previous research by Kjell Kleppe and Har Gobind Khorana. The polymerase chain reaction is a technique that allows scientists to make millions of copies of DNA in a short period of time. It is one of the most commonly used techniques in biotechnology. The polymerase chain reaction was responsible for the growth of forensic DNA analysis. Mullis shared the 1991 Nobel Prize in Chemistry with Michael Smith for their contributions to the development of DNA-based chemical methods. Mullis currently does independent research and gives lectures throughout the world.

Daniel Nathans

Daniel Nathans was born in Wilmington, Delaware, in 1928. He was the youngest of nine children born to Russian Jewish immigrant parents. The family had little money because his father was not employed for a long period of time after losing a family-owned business during the Depression. Nathans was motivated to achieve high goals in life because of his parent's high spirits in spite of their poverty. He attended Wilmington public schools and then studied chemistry, literature, and philosophy at the University of Delaware. Nathans was hoping to major in philosophy. However, his father encouraged Nathans to seek medicine as a career that would guarantee employment. Nathans was fortunate to receive a scholarship to study medicine at Washington University in St. Louis. While in medical school, Nathans was persuaded to follow a career in medical research. After an internship at Columbia-Presbyterian Medical Center in New York, Nathans accepted a clinical research position at the National Institutes of Health in Bethesda, Maryland. He then became a professor of microbiology at Johns Hopkins University School of Medicine in Baltimore, Maryland. Nathans performed collaborative research projects investigating the genetics of tumor formation. While doing this research he discovered restriction enzymes involved in cutting gaps into DNA. By using restriction enzymes scientists are able to insert new genes into an organism's DNA. This technique became a fundamental tool for genetic engineering research and helped create the field of contemporary biotechnology. Nathans shared the 1978 Nobel Prize in Physiology or Medicine with Werner Arber and Hamilton O.

Smith for the discovery and use of restriction enzymes. He died from leukemia in 1999.

Marshall Nirenberg

Marshall Warren Nirenberg was born in 1927 in New York City. Nirenberg's family moved to Orlando, Florida, when he was 10 years old. This is where he developed an appreciation for nature and planned on being a biologist. Nirenberg obtained a bachelor's and master's degree in biology from the University of Florida. He then went to the University of Michigan to work on a PhD in biochemistry. In 1960, Nirenberg was offered a research position at the National Institutes of Health in Bethesda, Maryland, to study protein synthesis experiments. DNA structure was worked out by the time Nirenberg studied genetics. However, little was known about DNA replication or gene function. One year later he discovered the way proteins are synthesized from DNA information. Nirenberg made this great discovery only 4 years after receiving his PhD. His enthusiasm for solving scientific problems was stated by Nirenberg's research supervisor Philip Leder. Leder commented, "Marshall was terrific. . . . That was enormously exciting . . . the way Marshall engaged the problem, and his enthusiasm and patience for very naive people like myself, was something that just excited and attracted me." Nirenberg shared the 1968 Nobel Prize in Physiology or Medicine with Robert W. Holley and Har Gobind Khorana for their interpretation and the deciphering of the role of genetic code and its function in protein synthesis. He received several national awards and honors for his original and later research endeavors. Nirenberg held the position of chief of biochemical genetics at the National Heart, Lung, and Blood Institute in Bethesda, Maryland, where he researches genetic disorders. He uses his appreciation of nature as an advocate for protecting the environment from human activities.

Severo Ochoa

Severo Ochoa was born in 1905 in Luarca, Spain. His mother took the family to Málaga, Spain after his father, an attorney, died when Ochoa was 7 years old. Ochoa became interested in biology while attending Málaga College. The writings of the Spanish neurologist, Ramón y Cajal, compelled Ochoa to seek a medical degree. Ochoa went to the Medical School of the University of Madrid and received a medical degree with honors. Upon graduation from medical school he developed an interest in teaching and doing medical research. Ochoa was awarded a fellowship to do biochemistry research at the Kaiser Wilhelm Institut für Medizinische in Germany and the National Institute for Medical Research in London, England. These experiences gave him the skills he needed to pursue new areas of research. He then held teaching positions at the universities of Madrid, Heidelberg, and at Oxford before coming to the United States to do research at New York University. His research at New York University led him to the discovery of an enzyme involved in the production of ribonucleic acid (RNA). This finding paved the way for more investigations into gene function. It is one of the fundamental principles of biotechnology. Ochoa shared the 1959 Nobel Prize in Physiology or Medicine 1959 with Arthur Kornberg for their discovery of the biological synthesis of genetic material. He was granted many honorary degrees and presented with several international awards for his scientific contributions. In addition, the Hospital Severo Ochoa in Madrid, Spain, was named in his honor. Ochoa also appears on a series of Spanish postage stamps. Ochoa returned to Spain as a science advisor. He died in Madrid in 1993.

Reiji and Tsuneko Okazaki

Reiji Okazaki was born around 1930 near Hiroshima City, Japan. He graduated with a PhD in genetics in 1953 from Nagoya University. Tsuneko Okazaki was born around 1933 in Central Japan. She is also known as Tuneko Okazaki. In 1956, she graduated with a PhD from the School of Science, Nagoya University. Little has been written about their lives. The Okazakis formed a research team at Nagoya University investigating the mechanism of DNA replication recently discovered by Arthur Kornberg. They uncovered a misunderstood feature of DNA replication that could not be explained with the research available at that time. The team noticed small pieces of DNA, which are now called Okazaki Fragments, which hinted to the full explanation of DNA replication. Their pioneering work was published in the *Proceedings of the National Academy of Sciences* in 1969. The paper was titled "Mechanism of DNA chain growth, IV. Direction of synthesis of T4 short DNA chains as revealed by exonucleolytic degradation." This research provided a fundamental understanding of DNA replication and is one of the foundations of biotechnology research. Reiji Okazaki died of leukemia in 1977 only a few years after the discovery. His disease was mostly likely due to radiation poisoning because he lived in Nagasaki when the second atomic bomb was dropped. Tsuneko Okazaki is currently a professor at Institute for Comprehensive Medical Science in the Fujita Health University in Japan. She continued to do cell biology research after the death of her husband. Both received honors for their contributions to genetics.

Richard Palmiter

Richard Palmiter was born in Poughkeepsie, NY, in 1942. He did his undergraduate studies in zoology at Duke University in Durham, North Carolina. Palmiter then went on to earn a PhD in biology from Stanford University in California. He continued doing research at Stanford University until taking a position at the University of Washington in 1974. In 1976, he accepted a concurrent appointment as an investigator with the Howard Hughes Medical Institute in Chevy Chase, Maryland. Palmiter is known as a versatile scientist and has made significant contributions to four different areas of molecular biology and animal physiology. Few scientists have this type of research expertise. He began doing research on the genetics of steroid hormone action. Later he focused his research efforts on the way certain proteins control the functioning of genes. This research provided much information needed for the production of properly functioning genetically modified organisms. Palmiter's biggest contribution to biotechnology was his revolutionary studies that produced transgenic mice. Transgenic mice are genetically altered by placing the genes of other organisms into their cells. This research opened the door for many biotechnology innovations including the use of gene therapy for correcting human genetic disorders. Currently, he is researching the genetics of human genetic disorders. Palmiter has received much recognition for his work and has been selected as a member of the prestigious National Academy of Sciences in 1984 for his research in biotechnology.

Louis Pasteur

Louis Pasteur was born in 1822 in Dole, France. He was the only son of a tanner who had little formal education. Pasteur admitted that he was not a serious student during his early education and preferred recreational activities and drawing. He entertained the idea of becoming an artist. However, his father discouraged that career path and wanted Pasteur to be a university professor. Pasteur attended primary and secondary schools in Arbois, France. The family then moved to Besançon, where Pasteur received two bachelor's degrees from Royal College in Besançon. He then went to a teaching college in Paris to do graduate work in the physical sciences and a PhD in chemistry. At the age of 26, Pasteur presented his first major discovery called chirality, which describes certain aspects of molecular shapes, to the Paris Academy of Sciences. This became a major contribution to biotechnology and is a fundamental principle of contemporary drug design. His research skills awarded him faculty positions at various French universities in Dijon, Strasbourg, Lille, and Paris. While at the University of Paris he focused his interest on biochemistry and made many discoveries in microbiology including the process of fermentation. This was another discovery that is currently exploited in modern biotechnology to produce a variety of chemicals and foods. Another major work, called Germ Theory, established the fact that microorganisms were responsible for infectious diseases. He has used this idea to find a method, now called pasteurization, to prevent the spread of disease from milk. Pasteur used the Germ Theory to develop a vaccination strategy for the fatal viral disease called rabies. He made many contributions to chemistry, biochemistry, immunology, and medicine. Recently, a science historian, Gerald Geison, discredited much of Pasteur's research claiming that it was based on plagiarized ideas. However, there is much debate about Geison's findings. In spite of these assertions, Pasteur was well known as a diligent and benevolent scientist. Pasteur's humanitarian view of science is echoed in his statement, "I beseech you to take interest in these sacred domains so expressively called laboratories. Ask that there be more and that they be adorned for these are the temples of the future, wealth and well-being. It is here that humanity will grow, strengthen and improve. Here, humanity will learn to read progress and individual harmony in the works of nature, while humanity's own works are all too often those of barbarism, fanaticism and destruction." Pasteur died in 1895 due to complications from a series of strokes. He received many awards and honors worldwide for his achievements.

Linus Pauling

Linus Carl Pauling was born in Portland, Oregon, in 1901. His father's occupation as a pharmacist very likely motivated Pauling's curiosity of science. He did his undergraduate education in chemical engineering from Oregon Agricultural College in Corvallis which is now Oregon State University. After teaching for a year at Oregon Agricultural College, Pauling worked on his PhD at California Institute of Technology in Pasadena, California. He studied chemistry, with minors in physics and mathematics. Pauling then spent a year studying the physics of atomic structure at various universities throughout Europe. He encountered many of the luminaries of atomic theory. This experience kindled his interest in analyzing the atomic structure of complex biological molecules using X-ray crystallography. Pauling was offered a position at the California Institute of Technology and remained there for rest of his professional career. He studied chemical bond structure for much of his time there. During World War II Pauling focused his research efforts on developing explosives, missile propellants, and gas detectors for the United States Navy. After the War, Pauling concentrated his efforts on determining the chemical bonds that make up proteins. His findings are a fundamental principle of biochemistry and have many important biotechnology applications. The 1954 Nobel Prize in Chemistry was awarded to Pauling for his research investigating chemical bonds and its application for determining the structure of biological molecules. His second son, Peter Pauling, worked in the same office area at Cambridge University with Francis Crick and James Watson. Peter Pauling prematurely released information to Crick and Watson about Linus Pauling's quest to investigate DNA structure. This compelled Crick and Watson to speed up their investigation into the structure of DNA. Linus Pauling also received the 1962 Nobel Peace Prize for his humanitarian work. Pauling credited his wife, Ava Helen Miller, with influencing his social consciousness. Linus Pauling was not afraid of controversy and regularly expressed unconventional scientific thoughts. He took moral positions about certain scientific matters and brought these issues out to the public. Pauling campaigned for many social causes and lent his expertise for international humanitarian concerns related to public health. His latest efforts promoted the health values of taking vitamin C to ward off disease. Pauling died at his ranch in Big Sur, California, in 1994.

Max Perutz

Max Ferdinand Perutz was born in 1914 in Vienna, Austria. He was born into a family of wealthy textile manufacturers who lost their businesses during the Nazi uprising in Germany. Perutz's parents wanted him to study law. However, he developed an interest in chemistry due the encouragement of a schoolmaster. This interest compelled him to obtain a degree in chemistry from the University of Vienna where he changed his focus to biochemistry. Perutz's new interest led him to Cambridge University in England for a PhD in biochemistry with funding from his family. He was able to apply his knowledge of chemistry to do X-ray crystallography on biological molecules. During his doctoral studies, Perutz's family became refugees when Germany occupied Austria. This almost prevented Perutz from finishing his education. His excellent research skills earned Perutz a fellowship to continue studying at Cambridge University where he also was offered employment. He worked there as a research assistant throughout World War II and then was put in charge as the head of the newly formed Medical Research

Council Unit for Molecular Biology at Cambridge University. Perutz's landmark work was the use of X-ray crystallography to determine the structure of the protein hemoglobin that carries oxygen in red blood cells. He developed a technique that revolutionized the study of protein structure and formed a major principle of biotechnology. Perutz shared the 1962 Nobel Prize in Chemistry with John Cowdery Kendrew for their studies on the structure of folded proteins. Perutz died of cancer in Cambridge, England, in 2002.

Stanley Prusiner

Stanley B. Prusiner was born in 1942 in Des Moines, Iowa. The family moved to Boston, Massachusetts, when Prusiner's father entered Naval officer training school. His mother then moved the family to Cincinnati, Ohio, to be near his father's family while the father was serving in the South Pacific. Prusiner said that Latin courses were a valuable part of his high school education. The courses helped him later when he was studying biology and preparing scientific publications. Prusiner was not motivated in high school, but still felt it important to attend college. He enrolled in the University of Pennsylvania to major in chemistry. The intellectual climate at the university inspired Prusiner to study in depth a variety of academic areas including philosophy and Russian history. He also became involved in extracurricular activities and sports. Prusiner then had the opportunity to assist in a research laboratory studying hypothermia. This produced an interest in seeking a medical career. Prusiner stayed at the University of Pennsylvania to complete his medical studies. Before completing his residency, Prusiner was offered a research position at the National Institutes of Health in Bethesda, Maryland. He then went to the University of California at San Francisco to do his residency. One of his first patients was suffering from Creutzfeldt–Jakob disease. It is a fatal degenerative disease of the nervous system that slowly destroys the brain. Prusiner was intrigued that Creutzfeldt–Jakob disease had an unknown cause and was similar to certain diseases found in animals. Upon investigating the disease he discovered it was caused by a self-replicating protein called a prion. Prusiner's prion stirred much debate in the scientific community because it was not believed that a simple protein could act like an infectious organism. The discovery of prions had a large impact on medical biotechnology. In 1997, Prusiner was awarded the Nobel Prize in Physiology or Medicine for his discovery of prions that led to a new biological principle of infection. He is currently with the Neuroscience Department at the University of California at San Francisco.

Steven Rosenberg

Steven A. Rosenberg was born in New York City in 1940. His keen sense of curiosity about nature and love of education steered him into premedical studies at Johns Hopkins University in Baltimore, Maryland. He stayed at the university to complete a medical degree. Rosenberg then completed his internship and residency at Peter Bent Brigham Hospital in Boston, Massachusetts. Rosenberg became more interested in conducting research than in doing clinical work. During his residency he went to Harvard University in Massachusetts to complete a PhD in biophysics. Upon graduation he worked as a research fellow at Harvard Medical School and then took a research position at the National Institutes of Health in Bethesda, Maryland. His greatest contribution to biotechnology was the use of genetically modified bloods to kill cancer cells. Another groundbreaking accomplishment was his collaboration with W. French Anderson in the first successful gene therapy trial on a human. His diverse education gave him the skills to create the genetically modified cells and conduct the experiment in a way that would not harm the subject. Aside from his work at the National Institutes of Health, Rosenberg serves as a professor of surgery at George Washington University, School of Medicine and Health Sciences in Washington, DC. He is a member of many prestigious medical and scientific organizations. In addition, Rosenberg received many national honors for his research on cancer and gene therapy.

Pierre Paul Emile Roux

Pierre Paul Emile Roux was born in 1853 in Confolens, France. His father was a schoolmaster who encouraged Roux to seek a higher education. Roux developed an interest in science while young and obtained a baccalaureate in sciences and a medical degree at the Medical School of the University of Clermont-Ferrand in France. His interest in medicine turned into a desire to perform medical research. To fulfill this desire he went on to pursue graduate work in chemistry. He enrolled in Sorbonne University and was recommended to work in Louis Pasteur's laboratory as a research assistant. Roux was considered the most distinguished of Louis Pasteur's students. Roux was described as possessing a clear, decisive, and analytical mind. He did not let the handicapping effects of his long-term respiratory disease stop him from putting all his energy into his scientific endeavors. His dissertation involved the work performed on developing Pasteur's rabies vaccine. Roux's reputation as a great researcher placed him in collaboration with many famous medical scientists in Europe. This gave him the scientific background to discover the role of bacterial toxins in causing disease. His findings are critical for carrying out the safe production of many biotechnology drugs. Roux collaborated on many medical studies that led to the development of modern biotechnology therapies. He co-founded the Pasteur Institute where he eventually became general director. Roux received many honors for his many medical and scientific contributions. He died in Paris in 1933.

Robert Rushmer

Robert Rushmer was born in 1914 in Ogden, Utah. He was raised in a family with scientific interests because his father was an optometrist and his grandfather was a physician. Rushmer followed in his grandfather's footsteps by obtaining a medical degree at Rush Medical College in Chicago after attending the University of Chicago. He then completed his residency at the Mayo Clinic in Rochester, Minnesota, with full intent of being a clinical physician. World War II changed Rushmer's work when he was assigned to conduct research on aviation medicine with the Army Air Corps at Randolph Field in Texas. This experience motivated him to seek a career in medical research. Upon leaving the Army, Rushmer accepted a position in physiology and biophysics at the University of Washington in Seattle. Rushmer carried out pioneering bioengineering research while at the University of Washington. His most notable medical achievements were the development of ultrasound for medical imaging and better tools for cardiac monitoring. Rushmer developed ways of monitoring the health of laboratory animals without having to remove body fluids or perform investigational surgeries. His instrument designs led to emerging techniques and technologies for distinguishing a variety of medical conditions. Many of his discoveries were developed into modern biotechnology instruments and practices. Rushmer received many awards and honors for his contributions to bioengineering. He died in 2001 in Redmond, Washington, after a long illness.

Frederick Sanger

Frederick Sanger was born in 1918 in Rendcombe, England. Sanger's awareness in science was nurtured by his father who was a physician. At an early age he became interested in biology and understood the importance of science and the scientific method in everyday life. Sanger was humble about his science career goals. In an interview he said, "I was probably above average but not an outstanding scholar. Initially I had intended to study medicine, but before going to University I had decided that I would be better suited to a career in which I could concentrate my activities and interests more on a single goal than appeared to be possible in my father's profession." He obtained an undergraduate degree and PhD in biochemistry at St John's College in Cambridge, England. Upon graduation he remained at the university to work at the Medical Research Council Laboratory of Molecular Biology. Sanger was one of the few people to be awarded two science Nobel Prizes. He received the 1958 Nobel Prize in Chemistry for his pioneering research on protein structure. His methodology helped build the foundations of modern protein biotechnology. In 1980, Sanger shared the Nobel Prize in Chemistry with Paul Berg and Walter Gilbert for their work on determining the amino acid sequences of DNA information. Sanger's later findings form the basic genetic principles used by all biotechnology applications. He has received many honors for his contributions to genetics and biotechnology.

Matthias Schleiden

Matthias Jakob Schleiden was born in Hamburg, Germany, in 1804. At first, he had little intention of becoming a scientist even though he was curious about the natural sciences. Instead, Schleiden obtained a law degree at Heidelberg and became a legal advocate in Hamburg, Germany. In 1831, because of his lack of success at practicing law, Schleiden studied botany and medicine at the University of Gottingen and Berlin University in Germany. Upon graduation he held professorships at the University of Jena in Germany and the University of Dorpat in Estonia. Schleiden was not satisfied with the traditional way in which plants were studied. He took advantage of the microscope to better understand plant function and structure. Little was speculated about function of plant cells since their discovery by Robert Hooke in 1655. Schleiden's ideas were contrary to many of the beliefs held by botanists of that time. However, his microscopic studies provided insights that were not investigated by his colleagues. In 1838, Schleiden published his book *Contributions of Phytogenesis* (Beiträge zur Phytogenesis) that detailed his microscopic observations of plants. Schleiden helped build the theory that the cell is the basic functional and structural unit of living organisms. This principle, called Cell Theory, is the foundation of biotechnology. Schleiden died in Frankfurt in 1881.

Theodor Schwann

Theodor Schwann was born in 1810 in Neuss, Prussia, which is now Germany. He attended the universities of Bonn, Warzburg, and Berlin in

Germany. Schwann's research emphasis in graduate school was animal physiology. His area of research in graduate school investigated the physiology of digestion. Schwann was the first to isolate enzymes involved in the digestion of proteins. After his graduate studies, Schwann was a professor of anatomy at the universities of Louvain and Liège in Belgium where he conducted much of his landmark research. Microscopy was still a new tool when Schwann was conducting studies on nervous system anatomy and development. In addition, little was known about the cellular composition of animals and research in animal development was in its infancy. Schwann was considered a master microscopist by most of his colleagues. He applied his expertise at the microscope to understanding embryological development. Schwann was the first to demonstrate that the mature tissues of all animals can be traced to embryonic cells. This finding is the foundation of stem cell research and forms the basic principle of developmental genetics. Schwann is sometimes called the father of cytology because he stressed the role of Cell Theory in explaining animal anatomy and physiology. Schwann extended the application of Cell Theory to animals in his book *Microscopic Researches into Accordance in the Structure and Growth of Animals and Plants.* This book supported the research findings of Matthias Schleiden. Schwann identified many types of cells making up the nervous system and had a cell called Schwann cell named in his honor. Schwann cells assist with nervous system function. Schwann's animal cell theory stimulated a great deal of research leading up to the birth of biotechnology. He died in 1882 in Cologne, Germany.

Maxine Singer

Maxine Frank Singer was born in New York City in 1931. Singer was raised in Brooklyn and attended New York City schools. She was encouraged to get a higher education because Singer's mother felt frustrated by her own lack of education. Singer became interested in science while attending Midwood High School in Brooklyn. After high school, she enrolled in Swarthmore College in Pennsylvania to study science. Singer became the first person in her family to go to college. She then continued her education at Yale University in Connecticut. Singer was awarded her PhD in biochemistry in 1957. Her first research position was at the National Institutes of Health in Bethesda, Maryland. It was at the National Institutes of Health that Singer did her groundbreaking research deciphering the genetic code with assistance of Marshall Nirenberg. This understanding of the genetic code was crucial for the further development of genetics needed for the birth of biotechnology.

Within a few years Singer was made chief of the Laboratory of Biochemistry at the National Cancer Institute in the National Institutes. Singer's research spanned many areas of biology and contributes much to the understanding of biochemistry and cell function. After serving at the National Cancer Institute she became president of the Carnegie Institute in Washington, DC. Singer has won many awards and honors for her distinguished career doing genetics research. She also has international recognition for public service related to genetics education and ethics. Singer retired from the Carnegie Institute in 2002 and remains an outspoken supporter of medical biotechnology applications for curing disease.

Lazzaro Spallanzani

Lazzaro Spallanzani was born in Scandiano, Italy, in 1729. Spallanzani at first did not pursue his interest in science as a career. He received a liberal arts education at the Jesuit College of Reggio and began to study law at the University of Bologna in Italy. His studies as a scientist began when he met a famous female professor, Laura Bassi, at the University of Bologna. She guided his studies in natural philosophy and mathematics. Upon graduation, Spallanzani became an ordained priest and accepted a professorship teaching Greek, logic, and metaphysics at the University of Reggio. He eventually accepted a position in natural history at the University of Pavia in northern Italy. Spallanzani made many contributions to biology that continue to have fundamental importance in biotechnology. His findings spurred many other scientists to reexamine older scientific beliefs that were inaccurate or not well tested. Spallanzani was the first to determine that fertilization occurs when semen contacts an egg. He also studied the digestive abilities of saliva. His most important contribution was his experiments supporting the biogenic principle of biology. This principle states that all organisms are produced from prior living organisms. This was contrary to the spontaneous generation view of John Needham, which followed the idea that living organisms are created from nonliving matter. His support of the biogenic principle permitted a better understanding of infectious disease. Spallanzani's great curiosity and unrelenting drive to understand nature by using the scientific method made him a well-recognized and respected scientist. He died in 1799 in Pavia, Italy.

Hermann Staudinger

Hermann Staudinger was born in 1881 in Worms, Germany. Education was important to his family. Staudinger's mother was a secondary school teacher and his father was Dr. Franz Staudinger who was a professor at the University of Applied Sciences in Worms. He graduated high school in Worms and studied chemistry at the universities of Halle, Darmstadt, and Munich in Germany. After doing postdoctoral work in Strasborg, he did research and taught at the Technical University of Karlsruhe, Eigenössische Technische Hochschule in Zurich, and at Albert Ludwigs University in Freiburg. He focused his research on the chemistry of organic polymers. He did some of his research in cooperation with BASF, a chemical company that produces plastics and other polymers. Staudinger's wife, Magda Woit, assisted with much of his research and was a co-author on many of his publications. Staudinger was known to be a prolific writer and authored many books on organic chemistry. The research findings he published in his books provided much insight in the polymers making up living organisms. He built the scientific principles needed to understand the chemistry of DNA and proteins. Many biotechnology products are polymers that were characterized by Staudinger. He was one of the developers of synthetic rubber and developed strategies for building other polymers. In 1953, Staudinger received the Nobel Prize in Chemistry for his discoveries in the field of polymer chemistry. Staudinger received many honors during his career and was noted as being the Father of Polymer Structure. After retiring at the age of 70, Staudinger stayed active in science by accepting a position as director of the Institute for Macromolecular Chemistry Baden-Württemberg. He died in 1965 in Freiburg.

Nettie Stevens

Nettie Maria Stevens was born in Cavendish, Vermont, in 1861. Her family moved to Westford, Vermont, where her father worked as a carpenter and a handyman. The parents encouraged education and had enough money to send the children to college. At first, Stevens did not seek a career in science. She worked as a librarian until she was 35 years old. Then she attended Stanford University in California to major in biology. Her professors were impressed with Stevens' excellent academic performance. As a result they recommended that she attend Leland Stanford University for a masters degree in biology. Stevens did her graduate work on the microscopic anatomy of new species of marine life. This training prepared Stevens for her future investigations of chromosomal function. Stevens then pursued a PhD in biology at Bryn Mawr College in Pennsylvania. At Bryn Mawr College she was fortunate to have Thomas Hunt Morgan as one of her professors. He stirred an interest in genetics in Stevens. She spent some time traveling through Europe and did a fellowship at the Zoological Institute at Würzburg, Germany. It was there that she started studying the role of chromosomes in inheritance. Upon graduation, Stevens was awarded an assistantship at the Carnegie Institute in Washington, DC. At Carnegie Institute, Stevens performed her revolutionary research that identified the role of chromosomes in sex determination. Her study was the first done on worms and insects. Later it was discovered that sex determination in humans followed the same principle. Steven's received much acclaim for her research and was well known as an astute scientist. She provided much of the foundation of modern genetic principles used in biotechnology. Unfortunately for Stevens, Edmund B. Wilson, who had read Stevens' research on chromosomes before publishing his own studies, was credited with a similar chromosomal inheritance theory. She received a Nobel Prize for the discovery. Stevens died of breast cancer in 1912.

Alfred Henry Sturtevant

Alfred Henry Sturtevant was born in 1891 in Jacksonville, Illinois. Sturtevant's father was a mathematics professor at Illinois College. However, when Sturtevant was 7 years old, his father moved the family to Alabama to do farming. Sturtevant attended public schools in Mobile, Alabama. He became interested in genetics at an early age and was known to draw up pedigree diagrams of his father's horses. Sturtevant pursued his undergraduate studies at Columbia University where he ultimately majored in biology. He conducted heredity studies as a student at the university and stayed to work on a PhD doing genetics research with Thomas Hunt Morgan. While working on his PhD, Sturtevant published the world's first genetic map using fruit flies as a model organism. His research began the race for genetic mapping that ultimately led to the Human Genome Project in 1989. Sturtevant was asked to stay at Columbia University where he did research in collaboration with Carnegie Institution in Washington, DC. He ultimately took a faculty position at the California Institute of Technology where he continued doing genetics research. Sturtevant explained that he found science exciting and rewarding. However, he enjoyed analyzing and explaining exceptions to established principles. He published many books on genetics and received national honors for his accomplishments. Sturtevant died in 1970 in Pasadena, California.

Walter Sutton

Walter Stanborough Sutton was born in 1877 in Utica, New York. When he was 10 years old his father, an attorney, moved the family to a ranch near Russell, Kansas. Sutton showed excellent mechanical skills operating and maintaining farm equipment, so he was encouraged to seek a career in engineering. After graduating from Russell High School, Sutton enrolled in the University of Kansas where he changed his interests from engineering to biology. His intense curiosity of biology led him to assist with histology research projects that encouraged him seek graduate work at the university. It was during his master's research that he discovered that the sperm and egg hold on to the individuality of their chromosomes during reproduction. From this information he hypothesized that chromosomes carry the genetic information from parents to offspring. At the age of 25, Sutton developed this landmark hypothesis that set the stage for genetics and biotechnology. He was the first to point out that chromosomes obeyed Mendel's rules of segregation. Sutton shared the fame for this idea with Theodor Boveri of Germany. He also learned at about the same time that chromosomes carried the genetic material. After presenting and publishing these findings he attended Columbia University in New York for a PhD in zoology. For no known reason, Sutton left his PhD studies to pursue a medical degree at the College of Physicians and Surgeons of Columbia University. Upon graduation, Sutton did his internship in surgery at Roosevelt Hospital in New York. Aside from being a practicing physician, Sutton conducted research at Columbia University. Sutton returned to Kansas and held a variety of agricultural and trade jobs not related to medicine. He then accepted a professorship of surgery at the University of Kansas School of Medicine. Again, Sutton carried out medical and research duties. Sutton died in Kansas in 1916 due to complications from a ruptured appendix. He made many revolutionary contributions to science before dying at the age of 39.

Wacław Szybalski

Wacław Szybalski was born 1921 in Lwów, Poland, which is now L'viv, Ukraine. Szybalski had a strong scientific upbringing. His mother, Michaline, had a doctorate in chemistry and his father, Stefan, was an engineer. After high school, he attended the Lwów Institute of Technology to major in chemistry. He was fortunate to finish his education at the time when Stalin's Soviet Army occupied Lwów after World War II and deported many of the professionals to the gulags of Siberia. Many of the college's professors were killed by the Nazis. Szybalski was able to complete the requirements for a chemical engineering degree and then moved around Poland, evading the Nazi and Soviet armies. He assisted the antiSoviet and the antiNazi underground until most of the postwar atrocities slowed down. This activity made if very difficult for him to complete his education without the risk of being arrested. Finally, he was able to earn a masters degree in chemical engineering at the Institute of Technology, Gliwice, Poland, and his Doctorate of Science in chemistry at the Gdansk Technical University in Poland. Szybalski did his research on the biochemistry of fermentation and conducted other studies on food chemistry. After completing his education he traveled throughout Europe and then to America moving his research from fermentation to genetics. Szybalski was an exceptional expert on fermentation and was responsible for many of the early visions of genetic engineering and other biotechnology techniques. Currently, he is Professor Emeritus of Oncology at the University of Wisconsin at Madison. He has many pioneering publications and received many honors for his accomplishments.

Howard Temin

Howard Martin Temin was born in 1934 in Philadelphia, Pennsylvania. His father was an attorney and his mother was noted being active in civic groups and educational organizations. Temin attended Philadelphia public schools and was fortunate to spend summers during high school at the Jackson Laboratory in Bar Harbor, Maine. This experience fueled an interest in biology and he attended Swarthmore College in Pennsylvania to major in biology. He then pursued graduate school for a PhD in embryology at the California Institute of Technology in Pasadena. Temin conducted his research in the laboratory of the famous geneticist Renato Dulbecco. It was in Dulbecco's laboratory that Temin developed a cell culture strategy for culturing retroviruses. This research made possible the first strategy for growing viruses related to the one that causes AIDS. Retroviruses are used in many types of genetic engineering techniques. Temin facilitated the growth of this area of biotechnology. After completing postdoctoral studies at the California Institute of Technology, Temin accepted a professorship at the McArdle Laboratory for Cancer Research associated with the University of Wisconsin at Madison. While working at the research center Temin discovered the enzyme reverse transcriptase that is needed for retrovirus reproduction. This finding is fundamental for understanding the ways to control retrovirus diseases. The enzyme is also useful in genetic engineering. Temin's research was honored when he shared the 1975 Nobel Prize in Physiology or Medicine with David Baltimore and Renato Dulbecco for their contributions to viral reproduction research. Temins continued at the University of Wisconsin studying cancer viruses. The notable geneticist David

Baltimore described Temin in these terms, "Howard displayed an unwavering commitment to the study of retroviral genetics and transformation of cells in culture during almost 40 years of research." Temin was known as an antismoking advocate, yet he died at the age of 59 in 1994 from lung cancer in spite of never smoking.

Arne Tiselius

Arne Wilhelm Kaurin Tiselius was born in 1902 in Stockholm, Sweden. His family moved to Gothenburg, Sweden, after his father died when Tiselius was very young. Tiselius attended a technical high school in Gothenburg and then enrolled in the University of Uppsala in Sweden to study chemistry. Upon graduation he worked with Nobel Prize winning chemist Theodor Svedberg who worked with chemical mixtures called colloids. Tiselius stayed at the University of Uppsala to complete a PhD in chemistry. His doctoral research involved finding ways to improve the separation of biological molecules. Tiselius' meticulous scientific studies earned him a position at the University of Uppsala where he remained for his professional career. He was most noted for producing the first refined electrophoresis equipment that was developed in 1937. Today, electrophoresis is most commonly used method for analyzing nucleic acids and proteins. Tiselius also developed new types of chromatography that greatly improved the separation of molecules important in biotechnology and medicine. In addition, he developed an adsorption detector that permits the differentiation and separation of carbohydrates, lipids, proteins, and nucleic acids. Being able to identify all of these molecules is critical for the growth and success of biotechnology. Tiselius received the 1948 Nobel Prize in Chemistry for his research on electrophoresis and adsorption analysis. He was also honored for creating an accurate method of determining blood proteins that is still used today for determining disease. Tiselius died in Uppsala in 1971.

Alexander Todd

Alexander Robertus Todd was born in Glasgow, Scotland, in 1907. He went to Allan Glen's School and then studied chemistry at Glasgow University. After gaining some research experience at the college he went to obtain a PhD in chemistry at the University of Frankfurt-on-Maine in Germany. Todd did his graduate research on the chemistry of bile which is secreted by the liver. He then went to Oxford University in England to complete another PhD. While at Oxford University, Todd worked on the chemistry of plant pigments with Nobel Prize winning chemist Robert

Robinson. Todd had several faculty and research appointments. He was at Edinburgh University in Scotland, the Lister Institute of Preventive Medicine in Chelsea in England, the University of London, and the University of Manchester in England. Todd then became a professor of organic chemistry at the University of Cambridge in England where he stayed. While at the University of Cambridge, Todd performed many research studies on biological molecules. These studies paved the way for the growth of modern genetics. Todd was awarded the 1957 Nobel Prize in Chemistry for his research on nucleotides metabolism. Todd's findings gave other scientists the background needed to formulate the structure of DNA and to work out the mechanisms of heredity. He was also awarded many other honors for his research. The significance of Todd's scientific accomplishments earned him the honor of being knighted in 1954 and was then elevated to Baron with the title Lord Todd of Trumpington (1962). Todd died in Cambridge in 1997.

Harold Varmus

Harold Elliot Varmus was born in Freeport, New York, in 1939. His family was among the Jewish immigrants from Eastern Europe. They highly valued education and encouraged Varmus to seek a higher education. Varmus' earlier years in Freeport public schools were not inspiring to him. However, he kept a close circle of friends who valued reading and various intellectual pursuits. His family then left New York because his father was transferred to an Air Force Hospital near Winter Park, Florida. At first, Varmus was planning a career in medicine. Then he changed his studies to poetry at Amherst College in Massachusetts and then pursued graduate work in English at Harvard. While at Harvard, Varmus decided again that he wanted to seek a medical career. He was then admitted to the Columbia University in New York and received his medical degree in 1966. Varmus held various positions in general practice and as a surgeon in the United States Public Health Service, a mission hospital in Bareilly, India, and Columbia-Presbyterian Hospital in New York. Varmus then became a clinical research associate at the National Institutes of Health in Bethesda, Maryland. His research involved the study of bacterial gene regulation. Varmus was fortunate to work with many notable geneticists at the National Institutes of Health. He then accepted a position at the University of California School of Medicine in San Francisco where he started researching retroviruses. This led to the discovery of cancer-causing genes called oncogenes. These genes are fundamentally important for biotechnology applications used to detect and treat cancers. Varmus shared the 1989 Nobel Prize in Physiology or

Medicine with J. Michael Bishop for their research on cancer-causing genes found in retroviruses. He currently serves as president of the Memorial Sloan-Kettering Cancer Center in New York City.

Craig Venter

John Craig Venter was born in 1946 in Salt Lake City, Utah. As a child Venter was bored with school and shunned memorizing what he considered to be useless facts. He started his higher education at the College of San Mateo in California and then enlisted in the navy during the Vietnam War. He served as medic in Vietnam. Venter said that his experience as a medic showed that by having the appropriate knowledge a person can save someone's life. At that point he developed an interest in science. After serving in the navy, he attended the University of California at San Diego to obtain an undergraduate degree in biochemistry and PhD in physiology and pharmacology. Venter became a professor at the State University of New York at Buffalo and then worked at the National Institutes of Health in Bethesda, Maryland. He left the National Institutes of Health when he became frustrated that his supervisors did not want to experiment with untested methods for rapidly sequencing the human genome. Venter formed a biotechnology company in California after leaving the National Institutes of Health. While Venter was serving as president of Celera Genomics he discovered a method of rapidly sequencing genomic information. Venter used the method to map the human genome before it was completed by the government funded Human Genome Project. His technique became a pioneering strategy used to investigate the genomic information of many organisms. This accomplishment raised some controversy because it placed government investigations in competition with corporate research. Venter felt that private companies are more efficient than large government projects developing biotechnology innovations. Many scientists were concerned that the information provided by private companies would not be available for further research projects without some type of cost. Venter currently directs the not-for-profit J. Craig Venter Institute in Rockville, Maryland. It reflects his interest in promoting genomics research for advances in public health and environmental concerns.

Rudolf Virchow

Rudolf Carl Virchow was born in Schivelbein, Poland, in 1821. Virchow went to the Friederich Wilheim Institute, a military academy in Berlin, where he obtained a degree in science. He then obtained a medical degree from the University of Berlin where he specialized in pathology. In medical school he was interested in the microscopic analysis of disease. Virchow then practiced medicine at Charite Hospital in Berlin. However, Virchow was relieved of his duties because his liberal views were contrary to the then German government. He then moved to teach anatomy at the University of Würzburg because of his political problems in Berlin. Virchow made many accomplishments at the university including forming a school of nursing. He then returned to Berlin to become chair of the Department of Pathology at the University of Berlin. Virchow was instrumental in training medical doctors to treat war injuries during the Franco-German war that took place in 1870–1871. His pathology research stressed on the newly formulated cell theory. In 1855, Virchow published his views about the cellular basis of disease in his book *Every Cell Stems from Another Cell* (omnis cellula e cellula). His belief that "disease is caused when cells revolt against the organism of which they are a part" became a fundament principle of heart disease and stress-related illnesses. However, Virchow did not agree with Pasteur's claim that certain diseases were caused by microorganisms. Virchow's contributions to cell theory and microscopic pathology provided many of the principles needed for the growth of biotechnology. He was known internationally as a noted scientist as well as a talented physician. The city of Berlin honored Virchow with a Monument Karlsplatz (Charles' Place) and the naming of Rudolf Virchow High School, Virchow Clinical Center, and Virchowstrasse (Virchow Street).

James Watson

James Dewey Watson was born in 1928 in Chicago, Illinois. Watson's childhood spent in urban Chicago did not prevent him from developing an interest in nature. His early interest in bird watching carried through to an intent on studying science while in high school. Watson's excellent performance in school earned him a scholarship to study zoology at the University of Chicago. He was 15 years old when he enrolled in college. Upon graduating from the University of Chicago, Watson received a graduate fellowship to do a PhD in zoology at Indiana University in Bloomington. He was refused admittance to the California Institute of Technology and Harvard University graduate science programs because his extensive background in the classics did not fit the goals of their zoology programs. Watson developed an interest in genetics while doing his PhD research in the laboratory of the Nobel Prize winning microbiologist Salvadore Luria. He subsequently applied to postdoctoral studies in microbiology in Copenhagen, Netherlands, as a Merck Fellow of the National Research Council. Watson then accepted a position with the

Cavendish Laboratory at the University of Cambridge, England. It was at University of Cambridge where Watson discovered the double helix structure of DNA with Francis Crick, Rosalind Franklin, and Maurice Wilkins. Watson returned to the United States to hold positions at the California Institute of Technology, Harvard University, and Cold Spring Harbor Laboratory in New York. With the onset of the Human Genome Project, Watson was made first director of the National Center for Human Genome Research at the National Institutes of Health in Bethesda, Maryland. He eventually returned to Cold Spring Harbor Laboratory to lead genetics research. Watson's discovery of DNA structure was awarded the 1962 Nobel Prize in Physiology or Medicine, which he shared with Francis Crick and Maurice Wilkins. He is the recipient of numerous international awards. Watson's contributions to Cold Spring Harbor Laboratory were honored with the naming of the Watson School of Biological Science at the Laboratory.

Maurice Wilkins

Maurice Hugh Frederick Wilkins was born in 1916 in Pongaroa, New Zealand. His father was an Irish immigrant physician who moved the family from New Zealand to Birmingham, England, when Wilkins was 6 years old. Wilkins attended King Edward's School and then studied physics at St. John's College in Cambridge, England. He then obtained a PhD in physics at the University of Birmingham in England. His dissertation research investigated the properties of luminescent chemicals. During World War II, Wilkins worked on a variety of projects related to radar and radioactive compounds. He assisted with the top secret Manhattan Project set up to develop the first atomic bomb. Wilkins was interviewed on radio and said that he was upset with the devastation caused by the bombs after they were dropped on Japan. After war, he taught at St. Andrews' University in Scotland and then moved to King's College in London, England, to work at newly formed Medical Research Council Biophysics Research Unit. His role at King's College was to interact with biologists to carry out biochemistry research. That is where he met Rosalind Franklin and started a collaboration with Francis Crick and James Watson at the Cavendish Laboratory in Cambridge, England. Wilkins' research contributed to the discovery of DNA structure. This research paved the way for modern genetics and biotechnology. He shared the 1962 Nobel Prize in Physiology or Medicine with Francis Crick and James Watson for their contributions to DNA research. Wilkins received many other honors for his scientific achievements. He died in London in 2004.

Ian Wilmut

Wilmut was born in Hampton, England, in 1944. As a child Wilmut hoped to join the navy to become a seaman. However, his color-blindness prohibited him from seeking this career path because he would not be able distinguish the colors of signal flags. In high school he developed an interest in science and was fascinated by the wool industries where he spent some of his youth in Yorkshire, England. Wilmut attended the University of Nottingham for his undergraduate work in agriculture. In an interview Wilmut commented that he could have easily stopped his education there and become a dairy farmer. However, his curiosity of biology compelled him to seek more education. Wilmut then went on to earn a PhD in animal genetic engineering from Darwin College of the University of Cambridge in England. His PhD research was on the freezing of boar semen. He also did experiments on frozen embryos and created the first calf produced from a frozen embryo in 1973. The animal was humorously named Frosty. This technique was of fundamental importance to cloning and stem cell research. It also became a regular procedure used for human fertility applications. His research also stirred much controversy that still exists today about the treatment of frozen human embryos. In 1974, Wilmut took a position at the Roslin Institute, an animal research breeding station in Scotland, where he still conducts research. It was at the Roslin Institute that he produced the first cloned sheep called Dolly. His technique differed from other cloning attempts because it used the DNA from an adult cell to produce the clone. Cloning before that time involved the separation of cells from a developing embryo. Wilmut has been recognized with many awards and honorary degrees for his groundbreaking research. He continues to write general reading and scientific books and articles on cloning.

GLOSSARY

Abiotic. Inanimate features of nature such as climate, rocks, and water.

Acetic acid. Another name for vinegar which is a fermentation product produced by fungi including yeast.

Adenosine triphosphate (ATP). A nucleic acid that stores and transfers energy within a cell.

Aerobic. The presence of oxygen.

Agarose. A sugar—like polymer used in certain types of electrophoresis gels.

Agriculture. The activities and technologies involved in growing organisms for commerce and food. Farming is the most common form of agriculture.

Allele. An alternative form of a gene.

Allergenomics. The study of the proteins involved in the immune response of animals and humans.

Allergic response. The characteristic signs of an allergy. It may include itching, a rash, sneezing, or watery eyes.

Allergy. An oversensitive immune response to a foreign substance in or on the body.

Amino acid. Biochemicals that are the building blocks of proteins.

Amino acid analyzer. An instrument that determines the types of amino acids of a protein.

Amino acid sequence. The order of the amino acids in a protein.

Amino acid sequencers. An instrument that determines the arrangement of amino acids of a protein.

Amplification. A method in which a specific DNA sequence is replicated in large amounts.

Amylase. An enzyme that breaks down starch.

Amyloid. A clump of prion proteins associated with diseased cells.

Anabolism. Chemical reactions that synthesize molecules for an organism.

Anaerobic. The absence of oxygen.

Anaphase. The third phase of mitosis.

Anaphase I. The third phase of meiosis I.

Anaphase II. The third phase of meiosis II.

Animal. An organism composed of many cells. Animals are characterized by movement, an obvious response to stimuli, and taking in food to fuel their metabolism.

Antibody. A protein used to fight infections produced by white blood cells called B—cells.

Anticodon. A region of the transfer RNA that binds to a specific codon sequence on the messenger RNA.

Antigen. A substance that the immune system uses to identify chemicals or organisms in the body.

Antioxidant. A groups of chemicals that slow down the natural decay of substances.

Antisense DNA. Refers to the strand of DNA that does not code for gene information.

Antisense RNA. A strand of RNA that does not code for gene information and binds to other RNA.

Apoptosis. A situation in which cells can program their own death using a strategy called programmed cell death.

Artificial insemination. A breeding technique in which selected sperms are introduced artificially into a female animal.

B-cell. A white blood cell that produces antibodies for the immune system.

Bacteria. A primitive single cell microorganism that feeds on dead matter or lives as a pathogen on animals and plants.

Bacteriophage. Common viruses that infect bacteria.

Balance. A device for accurately determining the mass of a chemical.

Behaviouromics. A study that maps the genetics behind the sum of ideas human beings can have relating to moral decision making.

Bibliomics. An investigation that applies high-quality and rare information, retrieved and organized by a systematic gathering of the scientific literature.

Biochemical. A complex molecule found mostly in organisms. It belongs to a group of chemicals called organic molecules.

Biodegradable. The ability for a chemical to rapidly break down on its own or due the action of microbes. Biodegradable materials are designed to break down into harmless chemicals.

Biodiversity. A measure of the number of types of organisms in an environment.

Bioenergy. Energy made available by the combustion of materials made from biological sources.

Bioenergetics. The chemistry and physics principles that govern the chemical reactions taking place in living organisms.

Bioengineering. The use of artificially derived tissues, organs, or organ components to replace parts of the body that are damaged, lost, or malfunctioning.

Biological molecules. A complex molecule found mostly in organisms. It belongs to a group of chemicals called organic molecules.

Bioinformatics. The collection, organization, and analysis of large amounts of biological data, using networks of computers and databases.

Biomics. The use of genomics, proteomics, and bioinformatics to develop a rational model for understanding the full functions of an organism's genetic material.

Bionanotechnology. The science of developing miniature computers and machines using biochemicals and components of organisms.

Biophysics. The application and understanding of physical principles to the study of the functions and structures of living organisms and the mechanics of life processes.

Bioprocessing. The use of uses cells, components of cells, or microorganisms to create commercially important products.

Bioreactor. A container for culturing microbes, growing cells, or carrying out chemical reactions used in biotechnology applications.

Bioremediation. Another name for environmental bioprocessing. A technique using cells, cell components, or organelles to break down wastes.

Biotechnology. Technologies that use living cells and/or biological molecules to solve problems and make useful products.

Biotransformation. Chemical modifications carried out by living organisms.

Blotting apparatus. An instrument for collecting certain types of DNA, RNA, or proteins in a concentrated sample.

Breeder. A person who breeds animal or plants for particular uses.

Breeding. The process of mating.

Calibration. The process of adjusting an instrument so that its readings are actually the value being measured.

Cancer. An abnormal spreading growth of cells.

Capillary. A very narrow hollow tube.

Carbohydrate. A group of biochemicals that compose sugars and starches. They are a major source of energy in organisms.

Cardiovascular system. The body parts that pump blood throughout the body. It is made up of the heart and blood vessels.

Catabolism. Chemical reactions responsible for the breakdown of molecule.

Cell. The smallest unit of life composing an organism.

Cell cycle. The events a cell goes through to carry out daily functions and the steps it takes to reproduce.

Cell membrane. A lipid and protein surface that covers the cell.

Cell theory or Cell doctrine. The assertion that all organisms are composed of cells.

Cellome. The entire accompaniment of molecules and their interactions within a cell.

Cellomics. The study of the genetics of cell functions.

Cellulase. An enzyme that breaks down the plant carbohydrate called cellulose found in paper and wood.

Centers for Disease Control (CDC). A government agency in Atlanta, Georgia, that is part of the United States Department of Health and Human Services. The CDC studies the spread of animal and human disease worldwide.

Central dogma. The principle that DNA programs for the protein and RNA that guides the synthesis of proteins.

Centrifugal force. A rotational movement that moves materials in solution away from the center of rotation.

Centrifuge. A machine that rapidly spins liquid samples and separates out various components of the sample by differences in their density.

Centriole. An organelle that assists the cell with reproduction.

Centromere. A region of the cell where doubled chromatids attach to one another.

Characterization. The identification of biological molecules.

Chiral. It means "mirror image". Some molecules exist having alternate shapes called chiral forms. The molecules are in effect mirror images of each other.

Chirality. The ability of a molecule to exist in two mirror image or chiral forms.

Chloroplast. A structure in plant cells containing the chlorophyll and carries out photosynthesis.

Cholesterol. A biochemical related to fat. It is necessary for body structure and function. However, too much cholesterol in the human diet is associated with cardiovascular disease.

CHO. An acronym for carbohydrates. The C stands for carbon, H for hydrogen, and O for the oxygen that makes up the chemistry of most carbohydrates.

CHOmics. The study of the role of carbohydrates in metabolomics.

Chromatin. The functional DNA in a cell.

Chromatinomics. The study of the chemistry controlling the genetic regulation of the functional DNA within a cell.

Chromatogram scanner. Also called a densitometer, this is an instrument designed to read the separation and intensity of bands on a chromatogram.

Chromatography. A chemical analysis technique that separates a mixture of chemicals into separate components.

Chromonomics. The study of gene location and arrangement on the chromosomes.

Chromosome. A strand of DNA that carries sets of genes and functions in the passing on of hereditary information.

Clone. A replicate of a cell or an organism.

Cloning. A technique for making multiple copies of an organism or a piece of DNA.

Coding strand. The DNA strand with the same sequence as the mRNA used for protein synthesis.

Codon. A triplet of nucleic acids in the genetic code that programs for an amino acid in a protein.

Complimentary DNA. A sequence of DNA, also called cDNA, produced from an RNA template.

Computational. Activities that involve mathematic calculation usually using computers.

Conjugation. A process in which two cells come in contact and exchange genetic material.

Consumer products industry. Any company that designs, manufactures, or markets items used everyday around the house such as appliances and clothing.

Cryopreservation. The process of storing biological samples or whole organisms at extremely low temperatures often for long periods of time.

Cytometer. An instrument for counting and sorting cells.

Cytometry. A method of sorting and counting cells.

Cytoplasm. The chemistry of the cell within the cell membrane and outside of the nucleus.

Cytosol. A gel-like fluid composing over half of the cells total volume.

Cytoskeleton. A meshwork of protein filaments in the cytoplasm giving the cell its shape and capacity for movement.

Disaccharide. Two different or similar monosaccharides bonded together.

Deletion mutation. The loss of nucleic acids from a DNA sequence.

Densitometer. Also called a chromatogram scanner, this is an instrument designed to read the separation and intensity of bands on a chromatogram.

Density. A measure of how heavy a solid, liquid, or gas is for its size or volume.

Developmental genomics. The study of the genetics of maturation and aging.

Deoxyribonucleic acid. Also known as DNA. It is the chemical information making up an organism's genetic material.

Diabetes. One of several diseases characterized by high blood sugar, excessive urination, and persistent thirst.

Diagnose. To determine the cause of such as in diagnosing a disease.

Differentiation. A process by which cells mature in order to carry out specific physiological tasks.

Diglyceride. A glyceride fat composed of two fatty acid chains attached to the glycerol.

Diploid. A genetic condition in which cells have a full set of genetic material consisting of paired chromosomes.

Disaccharide. A carbohydrate composed of two simple sugars bonded together.

DNA. An abbreviation for deoxyribonucleic acid. DNA is the chemical information making up an organism's genetic material.

DNA hybridization. A technique that binds specific segments of DNA to a strand of RNA.

DNA sequencer. An instrument used to determine the nucleic acid sequence of a length of DNA.

Dominant. An allele that determine the appearance of an organism.

Double helix. Two spiraling strands of nucleotides held together with chemical bonds that make up the DNA molecule.

Eastern blotting. A techniques used to collect and identify complex carbohydrates associated with cell structure.

Egg. The reproductive cell or gamete of females.

Electron. A minute electrically charged particle that orbits the nucleus of an atom.

Electrophoresis. A technique that uses electricity passing through a gel or narrow tube to separate biological molecules.

Electroporation. A technique that uses electricity for introducing DNA into a cell for genetic engineering.

Embryo. The early developing offspring stages taking place immediately after fertilization.

Embryogenomics. The study of genes that are involved in the development of organisms from the point of fertilization until birth.

Empiricism. A philosophy based on the principle that the only source of true knowledge is through experiment and observation.

Endomembrane system. A system of membranes in the cell made up of the nuclear envelope, endoplasmic reticulum, Golgi body, vesicle, and cell membrane.

Endoplasmic reticulum. The ER which is an organelle responsible for the production of most of a cell's protein and lipid components.

Endosymbiont. An organelle that is a prokaryotic organism living within the cells of another organism.

Energy. The ability of a chemical reaction to do work.

Environomics. A science investigating the role of the environmental on the expression of genetic material.

Enzyme. A protein that carries out a specific chemical reaction for an organism.

Enzymomics. A form of proteomics that investigates the function of enzymes.

Epigenetics. Changes in gene regulation and traits that occur without changes in the genes themselves.

Epigenomics. The study the changes in gene regulation and traits that occur without changes in the genes themselves.

Ergastoplasm. A system of sack-like membrane folds in areas where the endoplasmic reticulum is continuous with the plasma membrane.

Erythropoietin. A hormone in the body that stimulates the production of red blood cells.

Ethnicity. Similar organisms that have origins from different parts of the world.

Ethnogenomics. A science that evaluates the influence of ethnicity of the genomics of organisms.

Eugenics. The improvement of an organism by altering its genetic composition.

Evolution. A change in the traits of living organisms over generations.

Exons. The protein-coded DNA segments of a gene which remain following removal of introns.

Expression. This term is used synonymously with gene expression.

Extrachromosomal DNA. Genes not located on an organism's chromosomes.

Fat. A group of biochemicals composed mostly of carbon and hydrogen atoms. Most fats do not dissolve in water.

Fatty acid. A molecule consisting of carbon and hydrogen atoms bonded in a chainlike structure.

Ferment. To metabolism chemicals by fermentation.

Fermentation. An energy-producing anaerobic metabolism that converts sugars into other organic molecules.

Fertilization. The coming together of an egg and a sperm to from offspring.

Fertilizer. A substance used to help plants grow. It is usually used to supplement soil.

Fetus. A developing organism that is not capable of survival outside of the egg or the female body.

Filter matrix. A material used to separate solids from gases or liquids using filtration.

Filtration. A method of separating solids from gases or liquids by passing the mixture through one or more layers of a porous material called the filter matrix.

Flagella. Whip-like appendages found on some types of cells such as bacteria and sperm.

Flavr Savr tomato. A tomato that is genetically altered so that it ripens without softening.

Fluorescence. The ability for certain chemicals to glow or fluoresce when exposed to ultraviolet light.

Fluorescent in situ hybridization. A technique, also called FISH, for identifying particular genes on whole chromosomes using florescent DNA probes.

Fluid mosaic model. Describes the motion of the proteins in the membrane that appear arranged in a patchwork pattern.

Frameshift mutation. A change in the DNA that severely alters the genetic code.

Fructose. A common carbohydrate used for cell energy.

Functional group. The part of a molecule that provides molecules with their chemical and physical properties.

Fungus. A diverse group of organisms usually composed of cells formed into branched filaments. Fungi feed primarily on decaying matter. The plural of fungus is fungi.

G1. Gap 1 phase or the first phase of interphase.

G2. Gap 2 phase or the last phase of interphase.

G-force. A unit of force equal to the force exerted by gravity.

Gel reader. An instrument used to identify the separated molecules on an electrophoresis gel.

Gene. A segment of DNA that contains the genetic information for a particular trait.

Gene gun. A gun that shoots genes into a cell as a method for carrying out genetic engineering.

Gene expression. The process of a cell using a gene to produce trait characteristics.

Gene regulatory networks. GRNs are the on and off switches of genes.

Gene therapy. The modification of DNA by gene insertion to correct a genetic disease.

Genealogy. The investigation of a person's ancestry and family history.

Genetic engineering. Biotechnology techniques used to modify an organism's genetic material or to join together genetic material from one or more organisms to change an organism's characteristics.

Genetically modified organism. Any organism whose characteristics are changed using genetic engineering.

Genetic material. The hereditary material of an organism that programs for its traits.

Genetic pollution. The transmission of unnatural DNA from organism to another.

Genome. The complete genetic information making up an organism's genetic material.

Genomics. The study of genes and their function.

Gibbs free energy. The energy capable of doing work during a chemical reaction.

Glucose. A common carbohydrate used for cell energy.

Glyercide. Lipids composed of a fatty acid attached to a glycerol molecule.

Glycerol. A fat-like molecule that can bind to one, two, or three fatty acids.

Glycolysis. The oxidation of molecules to produce energy in the absence of oxygen.

GMO. An abbreviation for genetically modified organism.

Golgi body. An organelle responsible for modifying, storing, and shipping certain products from the ER.

Haploid. Cells that have one set of DNA.

Hemoglobin. A protein used to carry oxygen in the blood.

Heritable. Able to be passed along from one generation to the next.

High-growth industry. An industry identified as having much potential economic and employment growth.

HGT. An abbreviation for horizontal gene transmission.

Homogenous. The property of a mixture in which all the constituents are uniform throughout.

Homologous chromosomes. Chromosomes that pair during meiosis and represent one chromosome from each parent.

Horizontal gene transmission. The natural movement of DNA from one organism to an unrelated organism. The amount of DNA transferred programs for a functional gene. Also known as HGT.

Hormone. A group of biochemicals that work like chemical signals in organisms. Hormones control how cells work.

HOX gene. A regulatory sequence of DNA that controls body organization.

Human Genome Project. An international research effort to map and identify the human genome.

Hybrid. An organism produced by hybridization.

Hybridization. Adding new traits to an organism by breeding it with an unrelated organism or another related organism with different characteristics.

Hydrocarbons. Chemicals composed mostly of carbon and hydrogen. Fats are examples of hydrocarbons.

Hydrolysis. Chemical reactions that break down molecules using water.

Incubator. An instrument that maintains controlled environmental conditions needed to sustain the development or growth of cells, eggs, tissues, or whole organisms.

Insect. A group of arthropods characterized by three pairs of legs.

Insertion mutation. The addition of nucleic acids in a DNA sequence.

Intellectual property. A creation of the intellect that has commercial value including any original ideas, business methods, and industrial processes.

Interference RNA. RNAi is a method that modifies the function of mRNA as an attempt to regulate specific gene functions without altering the DNA or disrupting the function of other genes.

Interphase. The nondivisional stage of the cell cycle the prepares the cell for division.

Intron. Noncoding sequences of junk DNA interspersed among the protein-coding sequences in a gene.

Invertase. An enzyme that helps with the conversion of table sugar into fructose.

Inversion mutation. The rearrangement of a region of DNA on the chromosome so that its orientation is reversed with respect to the rest of the chromosome.

In vitro. Studies performed outside a living organism under lab conditions.

In vitro fertilization. An artificial breeding technique in which selected sperm and egg are united in a culture.

In vivo. Studies carried out inside living organisms.

Ion. An electrically charged particle.

Irrigation. To supply land with water usually for agricultural uses.

Isoelectric focusing. A modified form of electrophoresis used for distinguishing different types of proteins based on their response to pH.

Isolation. A method of separating a particular molecule from a mixture.

Isomer. An alternate arrangement or shape of a molecule.

Junk DNA. A common type of genetic information that either has no definitive role or helps in reducing the effects of environmental factors that damage DNA.

Karyotype. A photograph of chromosomes taken through a microscope.

Knockout. The deactivation of specific genes.

Laboratory information management systems. LIMS refers to the computers and software used to handle laboratory data.

Lactic acid. A fermentation waste produced by bacteria.

Leishmania. A protozoan that causes diseases in animals including humans.

Ligand. A chemical temporarily attached to particular molecules.

Ligase. An enzyme that connects two pieces of cut DNA strands.

Lipid. Any of a group of fat-like molecules that generally do not dissolve in water.

Lipidomics. The study of the functions of hundreds of distinct lipids in cells.

Liposomes. A spherical particle of lipid substance suspended in water.

Lyophilization. A process of rapidly freezing a solution of chemicals at low temperature followed by dehydration using a high vacuum.

Lyophilizer. An instruments for rapidly freezing a solution of chemicals at low temperature followed by dehydration using a high vacuum.

Lysosome. A vesicle responsible for recycling cell components.

M Phase. The mitosis phase of the cell cycle.

Mad cow disease. A brain disease of cattle and other diseases caused by an organism called a prion.

Macromolecule. Another name for a biochemical.

Mass. A measure of the amount of matter making up an object.

Medication. A chemical used for medical purposes or for treating diseases.

Meiosis. Sexual cell division that produces gametes.

Meiosis I. The first stage of meiosis.

Meiosis II. The second stage of meiosis.

Membrane diffusion. The movement of particles across a cell membrane from a more concentrated to a less concentrated area.

Messenger RNA. The mRNA is a nucleic acid derived from a copied segment of DNA during transcription.

Metabolic engineering. Altering a cell's DNA so that it carries out a desired metabolism. Metabolic engineering can be done on intact cells or cells used in bioprocessing.

Metabolism. The chemical reactions that carry out the living functions of an organism.

Metabolomics. The study of the gene expression that controls metabolism.

Metaphase. The second phase of mitosis.

Metaphase I. The second phase of meiosis I.

Metaphase II. The second phase of meiosis II.

Microarray. A method of studying how large numbers of genes interact with each other and respond to the cell's regulatory controls.

Microinjection. A technique for inserting DNA into a cell using a small capillary needle.

Microorganism. An organism that must be viewed through a microscope. Bacteria and fungi are examples of microorganisms.

Microplate reader. A special instrument designed to measure or monitor up to 96 chemical samples in a single procedure.

Microscope. An instrument that uses a combination of lenses or mirrors to produce magnified images of very small objects.

Microtome. An instrument that cuts thin slices of cells for observation through a microscope.

Missense mutation. A mutation that produces a genetic code change that alters a codon.

Mitochondria. An organelle or complex structure in cells that carries out aerobic metabolism.

Mitogen. A chemical that stimulates cell division.

Mitogenomics. A type of epigenomics that investigates the application of the complete mitochondrial genomic sequence.

Mitosis. Asexual cell division.

Mixer. An instrument for producing homogenous mixtures of liquids or solids.

Mobile phase. A liquid or a gas that pushes a mixture of chemicals being separated during chromatography.

Molecule. A chemical made up of two or more of the same or different atoms.

Monoglyceride. A glyceride fat composed of one fatty acid chain attached to the glycerol.

Monomer. A single molecular entity that may combine with other molecules to form more complex structures.

Monosaccharide. A single unit of a carbohydrate. It is also called a simple sugar.

mRNA. An RNA molecular, also called messenger RNA, that provides the amino acid sequence information for protein synthesis.

Multicellular organism. An organism generally composed of cells that carry out a specific set of tasks that contribute to the organism's survival.

Mutation. A change in the DNA information.

Nanotechnology. The science of developing miniature computers and machines.

National Aeronautics and Space Administration. NASA, a governmental organization involved in space flight and atmospheric research.

Nitrocellulose. A special type of paper used in blotting that attracts and binds to biological molecules.

Nondisjunctions. A type of aberration in which the chromosomes fail to successfully separate to opposite poles of the cell during division.

Nonsense mutation. Nucleotide changes that stop the synthesis of a protein before it is completely expressed.

Northern blotting. A technique used to collect and identify particular segments of RNA.

Nuclear magnetic resonance imaging. NMR or magnetic resonance imaging (MRI) is an analytical technique used to study the chemistry of animals, plants, and biotechnology products.

Nucleic acid. An organic chemical belonging to a complex group of molecules that form the genetic material and fuel cell functions.

Nucleoid. A region of a prokaryotes' cytoplasm where the genomic material is located.

Nucleus. The central structure of a eukaryotic cell containing the DNA.

Okazaki fragment. A small segment of newly copied DNA produced during DNA replication.

Organic molecule. Molecules composed of a carbon skeleton and arrangements of elements called functional groups.

Organelle. A small structure located in the cytoplasm of advanced cells. Organelles carry out specific functions of a cell.

Organism. A life form such as animals, bacteria, fungi, and plants.

Oxidation. A chemical reaction performed in aerobic respiration that combines oxygen with food molecules to cause a chemical change in which atoms lose electrons.

Paradigm. A philosophy of human thought.

Particle sizer. Also called a particle size analyzer, this is an instrument that measures the size of large chemicals and whole drugs.

Patent. Governmental approval that gives an inventor the exclusive right to make or sell an invention for a term of years.

Patenting. The act of applying for a patent.

Pathogen. An organism that causes disease.

Peptide. A chain of amino acids.

Pesticide. A chemical that kills organisms defined as pests and causes little harm to other creatures.

pH. A measure of the hydrogen ion concentration of a solution.

pH meter. An instrument that measures the hydrogen ion concentration of a solution.

Phagocytosis. The process a cell uses to take in large amounts of material from the environment.

Pharmaceutical. Any chemical used as medicine for treating a condition or illness.

Pharmacogenomics. The study of the relationship between a person's genetic makeup and their response to drug treatment.

Pharming. A term used to describe the use of GMOs for the production of medications.

Phospholipids. Glycerides that contain the element phosphorus.

Physiogenomics. The study of the genetics that explains the complete physiology of an organism, including all interacting metabolic pathways.

Physiomics. The study of the genetics of metabolic functions in the body.

Phytoremediation. The use of plants for bioremediation.

Pili. Plural form of the term pilus.

Pilus. A bacterial structure that can transfer DNA from one bacterium to another.

Pipette. A calibrated tube used for the delivery of a measured quantity of liquids.

Plant. A complex organism that carries out photosynthesis in chloroplasts.

Plasmid. A piece of DNA that exists independently of an organism's genomic DNA. They are mostly found in bacteria and are used in recombinant DNA research.

Polarized light. A beam of light in which the waves are all vibrating in one plane.

Polarimeter. Also called optical rotation instruments, this instrument measures the chirality of molecules.

Polyacrylamide. A polymer used in certain types of electrophoresis gels.

Polymer. Large biochemical formed by combining many smaller molecules or monomers into a regular pattern.

Polymerase. An enzyme that builds a copy of DNA or RNA.

Polymerase chain reaction. A technique, commonly called PCR, which is used to make multiple copies of DNA fragments.

Polysaccharides. A complex carbohydrate composed of chain of simple sugars.

Prenatal. The time before birth.

Primer. A nucleotide that attaches to DNA or RNA as a tool for replication of a section of DNA or RNA.

Prion. An infectious disease causing protein.

Probe. A molecule used to identify a particular sequence of DNA, RNA, or protein.

Promoter. A region of DNA which initiates transcription.

Protease. An enzyme that breaks down proteins.

Protein. A complex biochemical composed of chains of amino acids. Proteins provide function and structure for organisms.

Protein synthesis. The formation of protein by the cell using DNA information.

Proteomics. The study of gene expression and the protein composition of any organism.

Protista. A group of eukaryotes associated with disease.

Quantitative. Observations that involve measurements that have numeric values.

Raw material. Chemicals or substances used for manufacturing of commercial products.

Readout. A device on an instrument that displays a measurement.

Regulatory DNA. Chromosome segments and whole genes that function to regulate the expression of other genes.

Receptor. A protein on the cell surface that receives signals such as hormones from the environment.

Recombinant DNA. Genetically engineered DNA.

Recombination. The process of breaking and rejoining DNA strands to produce new combinations of genes.

Recyclable. Materials that are able to be reused without creating too much waste.

Red blood cells. Blood cells that carry oxygen from the lungs to the body. They are also called erythrocytes.

Reporter gene. A gene that identifies the expression of a particular portion of an expression vector.

Reproductively sterile. Not capable of producing offspring.

Resolution. The degree of detail used to characterize molecules.

Restriction enzyme. An enzyme that cuts DNA strands at specific locations.

Restriction fragment length polymorphism. Also called RFLP, these are variations occurring within particular sequences of DNA.

Retention. A measure of the rate at which a substance moves in a chemical separation system.

Retrovirus. A type of virus that has its genetic material in the form of RNA.

Reverse transcriptase. An enzyme the builds a DNA molecule from an RNA template.

Revolutions per minute. RPM refers to the number of times that a sample completes 360 degree rotation in one minute producing a measurement called g-force.

Rheometer. An instrument that determines the ability of a material to flow or be deformed.

Rheometry. The measurement of the ability of a material to flow or be deformed.

Ribonucleic acid. Also known as RNA. A chain of nucleic acids containing a ribose sugar component. They are involved in gene expression.

Ribosome. A structure found in the rough endoplasmic reticulum responsible for the manufacture of proteins.

Rickettsia. A disease causing microorganism classified with bacteria.

Rough endoplasmic reticulum. The RER which is a part of the ER responsible for manufacturing proteins.

Roundworms. Also known as nematodes. A simple cylindrical worm such as pinworms.

S phase. The synthesis or second phase of interphase responsible for DNA replication.

Selective breeding. The process of mating animal or plants to produce offspring with specific characteristics.

Semen. A fluid containing the sperm and nutrients to lubricate and nourish the sperm.

Shotgun method. A method of sequencing DNA that identifies randomly sequenced pieces of the genome.

Sialoglycoprotein. A protein that helps the body's immune system identify disease-causing organisms.

Sickle cell anemia. A genetic disorder that affects the hemoglobin of red blood cells. It causes problems with blood circulation.

Silent mutation. Mutations that cause genetic variation without changing the nature of the protein.

Single nucleotide polymorphisms. A SNP is a mutation of one base pair in a sequence of DNA.

Smooth endoplasm reticulum. The SER is a part of the ER that has a variety of functions including carbohydrate and lipid production.

Somatic cell. The name given to a body cell. These cells are not normally involved in animal reproduction.

Southern blotting. A technique to identify and locate particular genes in large segments of DNA.

Spectrophotometer. An instrument that uses light to identify and determine the concentration of a particular chemical in a solution.

Spectroscopy. A method that uses light to identify and determine the concentration of a particular chemical in a solution.

Sperm. A male gamete or reproductive cell.

Splice site mutation. Mutations that affect introns.

Spore. A structure that permits cells to evade damaging environmental changes that could dehydrate, freeze, or overheat active cells.

StarLink™ corn. A variety of genetically engineered corn carrying a bacterial gene that produces Bt toxin.

Stationary phase. Also called the immobile phase, it is a barrier that selectively slows down or accelerates the movement of different chemicals in the mixture being separated by chromatography.

Stem cell. An animal cell capable of developing into other types of cells or able to reproduce a whole organism.

Sterile. Referring to reproduction it means incapable of producing off-spring.

Sterol. A group of lipids similar to cholesterol that consist of a chain of carbons twisted into a pattern of rings.

Structural genes. DNA that carries the code for structural polypeptides and enzymes that build other structural components of a cell.

Superweed. A weed that cannot be controlled. They are accidentally pro-duced by overuse of pesticides and horizontal gene transmission from GM plants.

Surrogate. A female animal carrying offspring introduced by in vitro fertil-ization.

Synthesis. Chemical reactions that build molecules and polymers.

T-cells. White blood cells that assist the immune system in fighting infec-tion.

Technology transfer. The process of converting scientific findings from research laboratories into useful products by the commercial sector.

Telomere shortening. The process of a chromosome's telomeres becoming shorter after each round of mitosis.

Telomere. The end of a chromosome.

Telophase. The last phase of mitosis resulting in two similar cells.

Telophase I. The last phase of meiosis I.

Telophase II. The last phase of meiosis II resulting in the formation of gametes.

Telophase. The last phase of mitosis resulting in two similar cells.

Template. A molecule used to mold the structure or sequence of another molecule.

Terpene. A diverse group of complex fats that includes hormones, immune system chemicals, and vitamins.

Textile. Relating to material used to make fabric for carpeting, clothing, and other purposes.

Thermocycler. A laboratory instrument that repeatedly cycles through a series of temperature changes required for chemical reactions.

Thermodynamics. The relationships between heat and other physical properties such as atmospheric pressure, temperature.

Therapeutic. Refers to a treatment for curing or healing illness.

Tissue. A body structure composed of groups of cells. Tissues carry out specific functions in the body. For example, a large component of bones is composed of connective tissue.

Toxic. A chemical that poisons a cell or the body.

Trait. A distinguishing characteristic of an organism.

Transcription. The synthesis of an RNA molecule using a DNA template.

Transcriptomics. The study of the proteins that are made by the DNA at a particular time or under specific conditions.

Transduction. The transfer of genetic material from one cell to another using viruses as a vector.

Transfected. Cells that are genetically altered.

Transfer RNA. The tRNA carries amino acids used in the production of proteins by ribosomes.

Transgenic. A GMO that has DNA inserted from an unrelated organism.

Translation. The formation of a protein using mRNA as a template.

Translocations. The transfer of a piece of one chromosome to an unrelated chromosome.

Transposable element. A "jumping gene" or a piece of DNA that causes translocation.

Triglyceride. A glyceride fat composed of three fatty acid chains attached to the glycerol.

United Nations. An international organization that assists member countries with economic growth and social welfare.

United Nations World Health Organization. A division of the United Nations that works with health issues.

United States Bureau of Labor Statistics. A agency in the U.S. Department of Labor involved in collecting information on labor economics.

United States Department of Agriculture. A governmental organization that assists farmers and sets agricultural policies.

United States Department of Commerce. A governmental organization concerned with promoting economic growth.

United States Department of Labor. A governmental organization that sets workforce policies and assesses job forecasts.

United States Environmental Protection Agency. Also known as the U.S. EPA, this governmental organization works with environmental policy.

United States Food and Drug Administration. This government agency is a branch of the Department of Health and Human Services. They work with health and policy issues for cosmetics, drugs, food, and medical treatments.

Vacuole. A vesicle that is produced by the cell membrane.

Vaccine. A medication or treatment given to help the body defend itself from a particular disease.

Vector. A piece of DNA used for recombinant DNA technology. It is used to introduce DNA into the cells of animals, fungi, and plants.

Viroid. An infectious particle composed completely of a single piece of circular ribonucleic acid.

Virus. A small pathogenic organism. This can only reproduce within the cell of another organism.

Vitrification. A process in which cells are rapidly cooled in a manner that prevents ice formation in cells.

Vortex. A powerful circular current of water.

Water bath. A tub of water used to bathe a chemical reaction or culture of organisms used in industrial and laboratory procedures.

Water titrator. An instrument that determines the water content of a substance.

Weight. A measure of the force of atmospheric pressure and gravity on the mass of an object.

Western blotting. A technique used to collect a specific protein.

White blood cell. A type of blood cell that assists with immune system function.

World Intellectual Property Organization. An organization in Geneva, Switzerland, which promotes the protection of intellectual property throughout the world.

Yeast artificial chromosome. Also called a YAC, this is a vector made from yeast DNA used to insert particular genes into a cell.

Yield. An agricultural bioprocessing term referring to the amount of food or product obtained from an animal or plant. For example, the number of useable seeds produced by a field of soybean is its yield.

X-ray crystallography. A technique that uses X rays to determine the atomic structure of molecules that have been crystallized.

References and Resources

PRINT

Alberts, B., Lewis, J., Raff, M., Johnson, A., and Roberts, K. 2002. *Molecular Biology of the Cell.* London, UK: Taylor & Francis, Inc. This classic textbook provides detailed information about cell function and genetics. It is written for upper-level college students majoring in biological science fields. The book has detailed illustrations and in-depth descriptions of the metabolic pathways that are fundamental to an understanding of biotechnology principles. Many biotechnology applications related to genetics are discussed in the book.

Alcamo, I. E. 2000. *DNA Technology: The Awesome Skill.* Philadelphia, PA: Elsevier Science & Technology Books. The late Dr. Alcamo developed this book as a reader-friendly general education textbook to supplement biology and biotechnology coursework. It is written for high school and college students. The book provides basic information about genetics and DNA technologies. It supplies the reader with accurate explanations of the biotechnology procedures associated with genetic engineering. Easily to follow illustrations supplement the text.

Ausubel, F. M., Brent, R., Kingston, R. E., Moore, D. D., Seidman, J. G., Smith, J. A., and Struhl, K. 1989. *Current Protocols in Molecular Biology.* New York: John Wiley & Sons. This classic laboratory manual is a "how-to-do-it" guide for the basic techniques used in biotechnology applications. It covers the procedures for many genomic and proteomic investigations. The manual was developed for college-level biology and biochemistry courses and requires a basic understanding of biology, chemistry, and genetics. This book demonstrates the precision in laboratory skills needed to conduct biotechnology procedures.

Ausubel, F. M., Brent, R., Kingston, R. E., Moore, D. D., Seidman, J. G., Smith, J. A., and Struhl, K. 1989. *Short Protocols in Molecular Biology.* New York: John Wiley & Sons. This classic laboratory manual is an abbreviated guide for the fundamental experimental techniques used in basic biotechnology applications. It covers the basic procedures that form the basis of genomic and proteomic investigations. The manual was developed for college-level biology and biochemistry courses and requires a basic understanding of biology, chemistry, and genetics. This book demonstrates the critical types of laboratory skills needed to conduct biotechnology procedures.

Bains, W. 1998. *Biotechnology: From A to Z.* 2nd ed. New York: Oxford University Press. This basic book covers—defines and describes—many terms associated with biotechnology applications and research. It serves as a simple-to-understand handbook of the techniques common to genomics and proteomics. Most of the basic concepts needed to understand newsworthy biotechnology are covered in this book. Practical applications contemporary to the publication date of the book are mentioned for each technique described.

Barker, P. 1995. *Genetics and Society.* New York: The H. W. Wilson Company. This book investigates the societal issues associated with biotechnology applications and discoveries. Information about genetics has created much controversy in society dating back to the earliest discoveries of inheritance. The author provides an unbiased view of the various social concerns that result from genetic research. Many of the issues discussed in the book are still contemporary concerns raised by recent biotechnology developments.

Baxevanis, A. D., and Francis, O. B. F. (eds). 2004. *Bioinformatics: A Practical Guide to the Analysis of Genes and Proteins.* New York: John Wiley & Sons. This book is a technical compendium of bioinformatics applications. It is targeted at researchers interested in computer applications used for interpreting information gathered from genomic and proteomic studies. The book gives a good idea of the types of research being performed in bioinformatics. It also provides insights into future applications of bioinformatics.

Begemann, Brett D. 1997. Competitive Strategies of Biotechnology Firms: Implications for US Agriculture. *Journal of Agricultural and Applied Economics* 29:117–122. This journal article is written for owners of biotechnology firms interested in agricultural biotechnology. It provides information about current and future agricultural needs that can be solved with biotechnology developments. The article is written for a technical audience with a business or scientific background. However, it does provide the general reader good insight into the growth of agricultural biotechnology.

Bonnicksen, A. L. 2002. *Crafting a Cloning Policy: From Dolly to Stem Cells.* Washington, DC: Georgetown University Press. This well-written scholarly book provides a rational analysis of governmental regulations of cloning and other genetic technologies. The author discusses the importance of developing rational biotechnology policies that permit the growth of science while protecting the well-being of the public. It is a good book for understanding how science policy is determined.

Borem, A., Santos, F. R., and Bowen, D. E. 2003. *Understanding Biotechnology.* San Francisco, CA: Prentice Hall. This book is a simple-to-read introduction to biotechnology. It gives a brief history of major biotechnology events and an introduction to the fundamental principles of genetic engineering. Many of the critical areas in contemporary genomic research are discussed. It is written for a nontechnical audience wishing to understand everyday applications of biotechnology.

Bourgaize, D., Jewell, T. R., and Buiser, R. G. 1999. *Biotechnology: Demystifying the Concepts.* San Francisco, CA: Pearson Education. This basic biotechnology book investigates the ethical, political, and social issues raised in various fields of biotechnology. It covers the basic science needed to understand the biotechnology applications and procedures mentioned in the book. It is a well-written book that provides a comprehensive summary of biotechnology applicable to high school and college students.

Burrill, G. S., and Lee, K. B., Jr. 1991. *Biotech '92: Promise to Reality,* An Industry Annual Report. San Francisco, CA: Ernst & Young. This annual report written for business people shows the investment promises of early biotechnology applications. The report presents earlier technologies that led to the multitude of industries present today. It covers the rationale that investors use to develop various technologies. In addition, it gives good insight into the way scientific discoveries are used for industrial applications.

Butterfield, H. 1965. *The Origins of Modern Science 1300–1800.* London, UK: Free Press. This classical historical reference is an authoritative book on the early history of European science. It discusses the factors that led to the great discoveries following the Renaissance period. Many noted scientists who produced the foundations of biotechnology are discussed in detail. The book was written for a general audience as well as historians.

Chrispeels, M. J., and Sadava, D. E. 2002. *Plants, Genes, and Crop Biotechnology.* 2nd ed. Sudbury, MA: Jones and Bartlett Publishers. This college-level introductory biotechnology focuses on agricultural biotechnology related to crop production. It covers contemporary information on genomic and proteomic techniques used to improve crop plants. Basic botany principles provided to explain the biotechnology information are covered in the book. The book contains many useful illustrations that depict the biotechnology techniques discussed.

Cohen, D. 1998. *Cloning.* Brookfield, CT: Millbrook Press. This classical ethics book evaluates the social implications of biotechnology applications of cloning. The book is written for younger readers and discusses the history and current developments of cloning, gene therapy, and recombinant DNA technologies. It also examines the ethical ramifications of genomic procedures used on animals, humans, and plants.

Cohen, I. B. 1985. *Revolution in Science.* Cambridge, MA: Harvard University Press. In this book the author provides an understanding of the ways in which scientific revolutions are born. The book chronicles scientists who challenged the contemporary scientific ideologies to come up with many innovations that led to the field of biotechnology. This book is written for scholars and for general readers.

Conko, G. 2003. *Regulation: The Benefits of Biotech.* Washington, DC: Cato Institute. This report is prepared by the Cato Institute that promotes American public policy based on individual liberty, limited government, free markets, and peaceful international relations. The social benefits of modern developments in biotechnology are analyzed and discussed in this book. It was written for business people and policymakers.

Ellyn, Daugherty. 2006. *Biotechnology: Science for the New Millennium.* 1st ed. St. Paul, MN: EMC/Paradigm Publishing. This general biotechnology textbook is designed for high school and college level introductory biotechnology courses. It is simple to read and provides contemporary examples of each biotechnology topic covered. The book also provides a brief background of the biology and chemistry principles needed to understand the scientific principles of biotechnology. It is accompanied by a laboratory manual, which shows inexpensive laboratories that model biotechnology techniques.

Davis, J. 1990. *Mapping the Code: The Human Genome Project and the Choices of Modern Science*. New York: John Wiley & Sons. This book investigates the scientific discoveries and philosophies that led to the initiation of the Human Genome Project. It is written for a general audience and requires no scientific background to read. The book is a good historical reference of the period of time when the Human Genome Project was just getting underway.

DeGregori, T. R. 2003. *Bountiful Harvest: Technology, Food Safety and the Environment*. Washington, DC: Cato Institute. This report is produced by Cato Institute that promotes American public policy based on individual liberty, limited government, free markets, and peaceful international relations. It describes the economic and social benefits of agricultural biotechnology. The book discusses the benefits of modern biotechnology developments in agriculture. It was written for business people and policymakers.

Department of Energy. 1992. *Primer on Molecular Genetics*. Washington, DC: U.S. Department of Energy, Office of Energy Research and Office of Environmental Research. This government publication provides reading about the basic biology and chemistry needed to understand the Human Genome Project. The Department of Energy provided funding for the Human Genome Project and this book is part of their public education commitment to the project. It is written for a general audience.

Diamond, J. 1999. *Guns, Germs and Steel: The Fates of Human Societies*. New York: Norton, W. W. & Company, Inc. This fascinating book is a biologist's perspective on the effects of science and technology on human civilization. The author evaluates the social conditions that drive scientific advancements such as biotechnology. Also covered in this book are the impacts of technology on cultural attitudes and public health.

Drexler, K. E. 1996. *Engines of Creation: The Coming Era of Nanotechnology*. New York: Anchor Books. This book discusses the early investigations and inventions of nanotechnology. It provides good insight into the rationale used to make nanotechnology inventions. The book focuses on several of the biotechnology applications of nanotechnology. Many of the technologies mentioned in the book are now in development.

Durbin, P. T. (ed). 1980. *A Guide to the Culture of Science, Technology & Medicine*. London, UK: Free Press. This classical book evaluates the strategies scientists use when making new scientific discoveries. It helps explain why certain scientists are responsible for many of the great scientific theories and technologies. The book also looks into the driving force behind the growth of genetics and biotechnology.

Etherton, T. D., Bauman, D. E, Beattie, C. W., Bremel, R. D, Cromwell, G. L., Kapur, V., Varner, G., Wheeler, M. B., Wiedmann, M. 2003. *Biotechnology in Animal Agriculture: An Overview*. Ames IA: Council for Agricultural Science and Technology. This book written for the Council for Agricultural Science and Technology fulfills its mission to interpret and communicate scientific information about agricultural nationally and internationally. Covered in this book are current technological advances in biotechnology used in animal agriculture. The report was prepared for business people, policymakers, and the public.

European Commission. 1997. *Biotechnology (1992–1994)*. Final Report, Vol. 1. Brussels: Directorate-General Research. This report was prepared as an educational document for European government officials. It investigates and evaluates the status of biotechnology growth in between 1992 and 1994. This period represents a rapid growth of biotechnology industries in Europe and the United States. It is an informative document that provides an insight into the government's concerns on biotechnology.

European Commission. 1997. *Biotechnology: 1994–1998.* Progress Report 1997. Brussels: Directorate-General Research. This report was prepared as an educational document for European government officials. It investigates and evaluates the status of biotechnology growth in between 1994and 1998. This period represents a rapid growth and restructuring of biotechnology industries in Europe and the United States. It is an informative document that provides an insight into the government's concerns on biotechnology.

Federal Coordinating Council for Science, Engineering, and Technology. 1992. *Biotechnology for the 21st Century.* Washington, DC: U.S. Government Printing Office. This document by the Federal Coordinating Council for Science, Engineering, and Technology is meant to educate the public about developments in biotechnology. It was written during the rapid growth of biotechnology industries in the United States. It is simple to read and provides good insights into the early years of biotechnology.

Fransman, M., Junne, G., and Roobeek, A., eds. 1995. *The Biotechnology Revolution?* Oxford: Blackwell. This informative book assesses scientific advancements made in the earlier years of biotechnology. The authors evaluate the benefits and risks of biotechnology as a vehicle for technological change. It provides many critical views of the contributions biotechnology was making to agriculture, commerce, and medicine.

Fukuyama, F. 2002. *Our Posthuman Future: Consequences of the Biotechnology Revolution.* New York: Farrar, Straus and Giroux. This book is a good overview of the issues raised by biotechnology innovations. The author briefly describes various biotechnologies and then discusses their implications on future societies. It also evaluates the positive and negative global impacts of biotechnology. The book was written for general reading.

Furmento, M. 2003. *BioEvolution. How Biotechnology Is Changing the World.* San Francisco, CA: Encounter Books. This current general reading book provides a brief overview of medical advances made by biotechnology. The author looks at future scenarios that could result from continued growth of medical biotechnology applications. It takes a positive outlook of biotechnology. However, it also provides critical analyses of the implications of biotechnology medical advances.

Gallagher, W. 1996. *I.D.: How Heredity and Experience Make You Who You Are.* New York: Random House. This book is written as a genetics primer that provides background about DNA structure and function. It has good information for better understanding the applications and implications of genomic and proteomic research. The book was written for general reading and requires little prior knowledge of science.

Glazer, A. N., and Hiroshi, N. 1995. *Microbial Biotechnology.* New York: W.H. Freeman. This college textbook was written for biology students studying microbiology. The book provides an excellent overview of the scientific and technological principles needed to understand and develop biotechnology applications of microorganisms. It requires a background in general biology to understand much of the information provided in many of the chapters.

Glick, B. R., and Pasternak, I. (eds). 2002. *Molecular Biotechnology: Principles and Applications of Recombinant DNA.* Washington, DC: ASM Press. This upper-level college textbook was written for biology students studying genomics and proteomics. The book provides an excellent overview of the scientific and technological principles needed to understand genetic technologies. It requires background knowledge of general biology and genetics to understand much of the information provided in many of the chapters.

Good, M. L., Barton, J. K., Baum, R., and Peterson, I. (eds). 1988. *Biotechnology and Materials Science—Chemistry for the Future.* Washington, DC: American Chemical Society. This book written for the American Chemical Society is a technical overview of biotechnology advances in the materials sciences. It provides detailed information about the use of biotechnology to replace many chemical manufacturing processes currently used to make a variety of commercial chemicals. It is written for professionals and college students with a biology and chemistry background.

Grosveld, F., and Kollias, G. 1992. *Transgenic Animals.* San Diego, CA: Academic Press. This college-level book provides technical information about the production of transgenic animals. It provides good information about the earlier techniques used to genetically alter animals used in agriculture and in pharmaceuticals production. The book requires a background in biology and genetics to fully understand many of the concepts.

Harris, J. 1992. *Wonderwoman & Superman: The Ethics of Human Biotechnology.* Oxford: Oxford University Press. The book evaluates the ramifications of medical biotechnologies used to improve human health. It looks at the benefits and risks of using biotechnology to prevent and treat human disease. This book provides good insight into the early fears of biotechnology including issues related to improving the human race.

Herren, R. V. 2000. *The Science of Agriculture: A Biological Approach.* Stamford, CT: Delmar Learning. This college textbook focuses on the scientific principles of the agricultural industry. It includes many of the contemporary biotechnologies used in animal and plant agriculture. The book provides good background for gaining a better understanding of agricultural practices and technologies that pave the way for biotechnology innovations.

Jones, S. 1995. *The Language of Genes: Solving the Mysteries of Our Genetic Past, Present and Future.* New York: Anchor/Doubleday. The author provides introductory-level information about genetic studies used to study human origins. It provides useful information about the ways anthropologists use genomic information to understand human ancestry and evolution. The book also looks at the use of genetic explanations of language development and sexual behaviors. This book is written for general audiences.

Kay, L. *Who Wrote the Book of Life?* 2000. *A History of the Genetic Code.* Stanford, CA: Stanford University Press. This book details the history early genetics discoveries leading to the unraveling of DNA structure and function. It critically analyzes the variety of research studies that paved the way for modern genetics and ushered in biotechnology. It is a scholarly book written for a general audience.

Krimsky, S., and Shorett, P. (eds). 2005. *Rights and Liberties in the Biotech Age: Why We Need a Genetic Bill of Rights*. Lanham, MD: Rowman and Littlefield Publishers. This book challenges the seemingly unrelenting growth of biotechnology. The author proposes regulations that project society from ethical ramifications not addressed fully by the use of agricultural, commercial, and medical biotechnology applications. It provides a representative criticism of the social issues brought about by rapid technological change.

Kuhn, T. 1970. *The Structure of Scientific Revolutions*. 2nd ed. Chicago, IL: University of Chicago Press. This classical book is written from the perspective of a philosopher who studies the progress of science. The author explains the intuition that great scientists use when proposing radically new ideas in the sciences. Factors leading the growth of biotechnology are described in the book. It is a scholarly book written for general audiences.

Madigan, M. T., Martinko, J. M., and Parker, J. 2002. *Brock's Biology of Microorganisms*. 6th ed. Englewood Cliffs, NJ: Prentice Hall. This college textbook is designed for biology students studying microbiology. It provides background material for understanding the basic principles of biology and chemistry needed to understand biotechnology applications. The book also provides useful information about microorganisms needed for a full understanding of commercial and pharmaceutical biotechnology.

Micklos, D. A., and Freyer, G. A. 1990. *DNA Science: A First Course in Recombinant DNA Technology*. Burlington, NC: Cold Spring Harbor Laboratory Press and Carolina Biological Supply Company. This classical laboratory manual is designed for college students and technicians wishing to carry out biotechnology laboratory processes. It contains detailed information about the laboratory procedures commonly used in biotechnology research and applications. This book is written as a technical guide. However, it provides good insight into the complexity of biotechnology procedures.

Monod, J. 1974. *Chance and Necessity*. London, UK: Fontana/Collins. This classical book on the nature of genetic discoveries is written by one of the founders of gene function and structure. The author provides personal accounts of the scientists who conducted the original investigations forming the foundations of modern genetics and biotechnology.

Moore, J. R. (ed). 1989. *History, Humanity, and Evolution*. Cambridge, UK: Cambridge University Press. This scholarly book is a collection of thirteen original essays by foremost historians, philosophers, and scientists on the history of evolutionary thought. The book provides useful information about modern evolutionary thought that forms the basis of biotechnology investigations. It is written for a general audience and provides good insights into some of the ethical concerns associated with biotechnology applications.

Morris, J. 2005. *The Ethics of Biotechnology*. New York: Chelsea House Publishers. This simple-to-read book investigates the ethic issues associated with modern advances in biotechnology. It covers a full range of biotechnologies and provides a balanced view of the benefits and risks of biotechnology innovations. It is written for a general audience that does not have a science background.

Nasr, S. H. 1968. *Science and Civilization in Islam*. Cambridge, MA: Harvard University Press. The classical book investigates the role of science in Islamic thought. It is a scholarly book that critically analyzes the influence of Islamic thought on the growth of modern science worldwide. It also provides information about the influence of science on Islamic civilizations. Biotechnology innovations are mentioned along with other technologies.

National Cancer Institute. 1995. *Understanding Gene Testing*. Bethesda, MD: National Institutes of Health, Publication No. 96–3905. This public information document provides a brief summary of human gene testing. The simple-to-read publication is designed for readers with no scientific background. It has ample illustrations useful for promoting an understanding of the biotechnology used for human gene testing.

Rabinow, P. 1997. *Making PCR: A Story of Biotechnology*. Chicago, IL: University of Chicago Press. The anthropologist author investigates the corporate culture of Cetus Corporation during the creation of the polymerase chain reaction. It provides good insights into the people responsible for the global boom of biotechnology industries in the 1980s. This book is written for a general audience and does not require a science background to understand.

Ratner, M. A., Ratner, D., and Ratner, M. 2002. *Nanotechnology: A Gentle Introduction to the Next Big Idea*. Saddle River, NJ: Pearson Education Inc. This simple-to-read nontechnical book guides the reader through the science and applications of nanotechnology. The author investigates the current technologies and prognosticates future directions for nanotechnology research. Many biotechnology applications of nanotechnology are mentioned. It is useful information for readers wishing to know the basic principles of bionanotechnology.

Rogers, M. 1977. *Biohazard*. New York: Alfred A. Knopf. This classical book is one of the first rational critiques of the growth of modern biotechnology. The author focuses on the first Asilomar conference held in 1975 to discuss the implications of genetic engineering. It provides insights into the thoughts of scientists who were responsible for the birth of biotechnology. It is written for a general audience and provides basic information about the science of genetic engineering.

Scheppler, J. A., Cassin, P. E., and Gambler, R. M. 2000. *Biotechnology Explorations: Applying the Fundamentals*. Washington, DC: American Society of Microbiology. This college-level book is designed to give the reader a general understanding of the biology and chemistry needed to understand biotechnology. It was written to reinforce learning by providing many opportunities to investigate biotechnology applications. It is a good general reference book for readers wanting details about the scope of biotechnology.

Schrodinger, E. 1967. *What Is Life?* Cambridge, UK: University of Cambridge Press. This classical scholarly book gives a biologist's perspective of the characteristics attributed to living organisms. It investigates the principles of biology and chemistry necessary for understanding how organisms surface. The book is interesting reading for readers who want insight into how genetic modifications could influence the survival of humans and other life forms.

Shmaefsky, B. R. 2005. *Biotechnology on the Farm and Factory*. New York: Chelsea House Publishers. This simple-to-read book is useful for gaining a general understanding of agricultural and industrial biotechnology. It covers the full scope of modern biotechnology applications used in agriculture and in the industrial manufacturing of foods and commercial products. The biology and chemistry principles needed to understand biotechnology are briefly covered.

Thieman, W. J., Palladamo, M. A., and Thieman, W. 2003. *Introduction to Biotechnology.* San Francisco, CA: Benjamin Cummings. This college textbook was written to gives detailed basic information about bioinformatics, genomics, and proteomics. It is applicable to science majors and readers wishing to know the fundamental science of biotechnology. Ethical considerations of biotechnology innovations and applications are also discussed.

Thompson, L. 1994. *Correcting the Code: Inventing the Genetic Cure for the Human Body.* New York: Simon and Schuster. The simple-to-read book investigates the science and technology of gene therapy. It discusses the biological principles behind the genetic engineering procedures used to cure or treat human genetic disorders. This book was written during the first successful trails of gene therapy on humans. It was written for a general audience.

Watson, J. D., Levine, M., Losick, R., and Baker, B. 2003. *Molecular Biology of the Gene.* San Francisco, CA: Benjamin Cummings. This classical book is the latest edition of the most comprehensive college textbook about molecular genetics. The main author, James D. Watson who co-discovered DNA structure, provides detailed information about the biology and chemistry of genomics and proteomics. This book requires a background in biology and chemistry to understand most of the concepts covered. It is useful book for gaining a full understanding of biotechnology.

Winston, M. L. 2004. *Travels in the Genetically Modified Zone.* Cambridge, MA: Harvard University. This well-written book provides a scientist's perspective of the pros and cons of agricultural biotechnology. It gives a balanced assessment of the benefits of risks of genetically modified crops and livestock. The book addressed the environmental issues as well as the impact of agricultural biotechnology on public health.

WEB

Access Excellence: http://www.accessexcellence.org. This educational Web site maintained by the National Health Museum provides basic information and updated news about biotechnology. It also has many useful resources for teachers.

Action Bioscience: http://www.actionbioscience.org. This valuable Web site maintained by the American Institute of Biological Sciences provides many teaching resources on a variety of biotechnology issues. It also has useful general reading materials.

BioABACUS: http://darwin.nmsu.edu/~molbio/bioABACUShome.htm. This Web site maintained by New Mexico State University is a searchable database of abbreviations and acronyms used in biotechnology. It describes a variety of terms commonly used in biotechnology research and applications.

BioCom: http://www.bio.com/. This commercial Web site provides valuable industry and research news about recent biotechnology developments having commercial and medical value. It also gives information about the variety of biotechnology careers currently available worldwide.

BioTech: Life Sciences Resources and Reference Tools: http://biotech.icmb.utexas. edu. This Web site maintained by the University of Texas is designed to the foster public's knowledge of biology and chemistry principles. Much of the information provided on the Web site is applicable to biotechnology.

Biotechnology Industries Organization: http://www.bio.org/. This commercial organization Web site is designed to share information between biotechnology companies and researchers. It provides a wealth of information about new biotechnology developments and trends.

Biotechnology Institute: http://www.biotechinstitute.org. This educational Web site provides a wealth of biotechnology resources for the public and for teachers. It has links to a variety of Web sites that present information about biotechnology advances and careers.

Cambridge Healthcare Institutes: http://www.genomicglossaries.com/. The commercial Web site provides simple-to-understand definitions about genomics terms.

Cold Spring Harbor Laboratory: http://www.cshl.org. This classic Web site houses much of the United States' historical information of genetics and genomic biotechnology. It has resources for the public and for educators.

Council for Biotechnology Information: http://www.whybiotech.com. The commercial Web site has many reliable resources that explain various concepts of biotechnology. Much of the information is written for the public.

European Federation of Biotechnology: http://efbpublic.org/. This professional organization Web site provides a variety of valuable information about biotechnology initiatives throughout Europe. Information about biotechnology policy applicable to global trade is available on this Web site.

Food and Agriculture Organization of the United Nations: http://www.fao.org/. This Web site maintained by the United Nations provides a wealth of information about global attempts to reduce famine using new developments in agriculture. Many issues related to biotechnology are addressed on the Web site.

Food and Nutrition Information Center: http://nal.usda.gov/fnic/. This governmental Web site has general information for the public about agriculture and nutrition. It has many references to biotechnology applications related to agriculture and food safety.

Human Genome Project: http://www.ornl.gov/sci/techresources/Human_ Genome/home.shtml. This governmental Web site was designed to share information about the Human Genome Project and other genomic studies. It has sections designed for the public and for educators.

Information Systems for Biotechnology: http://www.isb.vt.edu. This Web site maintained by Virginia Tech University has a wealth of research information about advances in agricultural biotechnology. A valuable on-line newsletter is available on the Web site.

National Center for Biotechnology Information: http://www.ncbi.nlm.nih.gov. This governmental Web sites is wholly designed to educate the public and educators about the science and applications of biotechnology. It links to many other useful biotechnology resources.

Pedro's BioMolecular Research Tools: http://www.public.iastate.edu/~pedro/ research_tools.html. This Web site was developed for researchers seeking technical information about biology and chemistry databases needed for doing biotechnology research and development. It has many interesting chemical modeling resources used to study proteomics.

Pew Institute on Food and Biotechnology: http://pewagbiotech.org. The organization provides unbiased information about advances in agricultural biotechnology. It is designed to educate policymakers and the public.

Transgenic Crops: An Introduction and Resource Guide: http://www.colorstate.edu/programs/lifesciences/TransgenicCrops. This Web site maintained by Colorado State University was designed to educate the public about agricultural biotechnology related to crop production. It has many useful resources for educators.

INDEX

About the Author

BRIAN ROBERT SHMAEFSKY is a professor of biology at Kingwood College in Texas. He was an industrial chemist in the biotechnology industry before becoming a college professor who teaches biology and biotechnology. Dr. Shmaefsky studied at Brooklyn College in New York and did his graduate work at Southern Illinois University and the University of Illinois. He continues to consult in the biotechnology industry and trains teachers to perform biotechnology in their classrooms.